T0226248

Synthesis Lectures on Mathematics & Statistics

Series Editor

Steven G. Krantz, Department of Mathematics, Washington University, Saint Louis, MO, USA

This series includes titles in applied mathematics and statistics for cross-disciplinary STEM professionals, educators, researchers, and students. The series focuses on new and traditional techniques to develop mathematical knowledge and skills, an understanding of core mathematical reasoning, and the ability to utilize data in specific applications.

Eric Stade · Elisabeth Stade

Calculus: A Modeling and Computational Thinking Approach

 Springer

Eric Stade
University of Colorado Boulder
Boulder, CO, USA

Elisabeth Stade
University of Colorado Boulder
Boulder, CO, USA

ISSN 1938-1743   ISSN 1938-1751   (electronic)
Synthesis Lectures on Mathematics & Statistics
ISBN 978-3-031-24683-8   ISBN 978-3-031-24681-4   (eBook)
https://doi.org/10.1007/978-3-031-24681-4

This Springer imprint is published by the registered company Springer Nature Switzerland AG
The registered company address is: Gewerbestrasse 11, 6330 Cham, Switzerland

Preface

This first-semester calculus text represents a reimagining of (roughly the first half of) the revolutionary textbook *Calculus in Context* by James Callahan, David A. Cox, Kenneth R. Hoffman, Donal O'Shea, Harriet Pollatsek, and Lester Senechal, of the Five College Calculus Project. Many of the big ideas, and much of the content, of the present textbook derive, in modified form, from that one.

In both this book and *Calculus in Context*, the iconic "*SIR*" (for "Susceptible/Infected/Recovered") epidemic model is introduced at the very beginning. This model is developed from intuitive ideas, without requiring any of the foundational Calculus notions—limits, continuity, and so on—that often populate the first week or so of a beginning course in Calculus. Instead, those foundational notions are explored, here, only once the *SIR* equations have prompted a closer look at some of the mathematical underpinnings. In this way, the motivation—or at least a motivation—for studying calculus is situated front and center.

In the *SIR* scenario, it's not possible to quantify the populations of susceptible, infected, and recovered individuals with nice, compact formulas. Instead, numerical methods, using computers, are required to analyze these populations. Thus, in a traditional calculus course, where simple formulas are typically foregrounded, and numerical methods are generally eschewed, the *SIR* model cannot be studied in real depth. By contrast, here (as in *Calculus in Context*), we embrace numerical methods! Indeed, we see *SIR* as an excellent "excuse" to acquire some basic programming notions and skills. (All such skills and notions are developed along the way; none are required at the outset.)

And the programming, in turn, helps to elucidate the calculus. More specifically, the concept of a rate of change, which is perhaps *the* central idea of Calculus, is also at the heart of the "Euler's method" algorithm that we use to solve the *SIR* equations. Familiarity with Euler's method then promotes facility with rates of change.

Programming also requires certain linear, "computational" ways of thinking. Thus programming fulfills a role, in this book, that rigorous, formal proofs have sometimes played in earlier calculus texts. Of course, there's nothing wrong with rigorous, formal proofs. They can be powerful things of beauty. But programming is a skill that has perhaps wider, more direct utility in life beyond Calculus.

All of the above perhaps suggests that this text is well off of the beaten path. This may be true in some respects. But this book does cover most of the "standard" calculus notions: derivatives; differentiation rules and formulas, linear, polynomial, trigonometric, exponential, and logarithmic functions; inverse functions; optimization; Riemann sums; antiderivatives; integration; the Fundamental Theorem of Calculus; integration by substitution; separation of variables; and so on. The approach is different, though: it's contextual, and it's computational.

A number of the contexts considered in this book arise from the Life Sciences. In addition to the *SIR* model, we investigate (single and dual species) population growth, circadian rhythms, neural impulses, and several other biological phenomena. However, this text is not intended (solely) as a Calculus for Life Sciences textbook. Other contexts— projectile motion, work and force, power and energy, radioactive decay, solar energy, and so on—are explored as well. And again, the Calculus notions developed here are universal. Our intention is that this book be suitable for a wide variety of first-semester Calculus courses.

As noted above, this book owes a tremendous debt of gratitude to the Five College Calculus Project, and to James Callahan, David A. Cox, Kenneth R. Hoffman, Donal O'Shea, Harriet Pollatsek, and Lester Senechal, authors of *Calculus in Context*. All of the Five College folks have been incredibly generous and gracious in letting us use their material freely. We thank them not only for this material, but for their groundbreaking approach to Calculus.

Further, we thank them for allowing us the freedom to reimagine and update their text using our own voices, ideas, perspectives, and experiences in the classroom. Theirs is a difficult book to improve upon, but we did try to make the present text our own. Key new features of our textbook include the following:

- Additional topics related to epidemics, including familiar, COVID-age notions of reproduction number and herd immunity;
- Programming exercises and examples based on computers, and on the free, web-based, Sage (also known as Sagemath) computer mathematics package, instead of programmable calculators;
- Additional scaffolding, in the form of numerous new worked examples;
- A variety of new contexts, including an Ebola outbreak, circadian rhythms, the genetic toggle switch, diffusion across a membrane, neural impulses, the Colorado flood, solar energy, and so on;
- Selection and ordering of topics that is somewhat more in line with "traditional" Calculus I courses;
- A number of additional exercises, from "drill" to abstract.

We are particularly grateful to James Callahan, who was our primary liaison to the Five College Calculus Project, and who engaged in many conversations and email exchanges with us. These communications with Jim made the writing of this text so much easier and made the end product so much better.

We also thank a number of University of Colorado faculty, including David Webb of the School of Education, Robin Dowell and Mike Klymkowsky of the Department of Molecular, Cellular, and Developmental Biology, and Richard Holley (Emeritus) of the Department of Mathematics, for the various ways in which they helped to shape this text.

Additionally, deep, deep gratitude goes out to all of the students who influenced this text. This includes all of the graduate student instructors and Teaching Assistants, as well as the undergraduate Learning Assistants, with whom we have taught a course based on this text. It also includes the hundreds of students who have taken that course. All of these students have had profound impacts on both the content and the pedagogy behind this text.

Finally, we thank our extraordinary sons Jack and Nick, for enduring everything that went into the writing of this text, but primarily just for being Jack and Nick.

Boulder, USA   Eric Stade
                                                                                                                     Elisabeth Stade

Contents

A Context for Calculus

1.1   Introduction: Calculus and Prediction

Calculus is often described as the mathematics of *change*. And one of the most significant things we can do about change is predict it.

If we can predict the way a situation will evolve, then we can prepare for it. Perhaps more critically: if we can predict how the evolution of a phenomenon will depend on present, or imposed, conditions, then we can endeavor to establish the conditions that will yield the most desirable outcome. So prediction is a good thing to be able to do.

For these reasons, we like to think of calculus as the mathematics of *prediction*. And all of calculus rests, at some level, on the following simple *prediction principle:*

> **If you know how fast you're going, then you know how far you'll get in a given amount of time.**

The prediction principle

Now this principle, as stated, may read like a statement about moving objects—about position, displacement, velocity, and so on. And certainly this principle does apply to all of those things. Indeed, Calculus was invented (discovered?) in the mid-to-late 1600s, by Isaac Newton and Gottfried Leibniz (more or less independently), in part to address problems in planetary motion.

But the principle has much, much broader relevance. It applies in any situation where you know the rate at which things are changing (that is, you know "how fast you're going"). It applies, then, to many, many, many phenomena—populations, epidemics, temperature, climate, neural impulses, protein concentrations, chemical reactions, radioactive decay, etcetera, etcetera—that evolve with respect to time (or that vary spatially, or depend on some other dimension or dimensions).

© The Author(s), under exclusive license to Springer Nature Switzerland AG 2023   1
E. Stade and E. Stade, *Calculus: A Modeling and Computational Thinking Approach*,
Synthesis Lectures on Mathematics & Statistics,
https://doi.org/10.1007/978-3-031-24681-4_1

Of course, since calculus is the mathematics of prediction, it *is* mathematics, among other things. Therefore, to apply calculus—and the above principle—to the study of an evolving phenomenon, we first need to describe that phenomenon in mathematical terms. Doing so is called **(mathematical) modeling**. The art and science of mathematical modeling will constitute major emphases of this book.

Models that describe, mathematically, how fast you're going are called **dynamical systems**. And once we have a dynamical system in hand, a major goal, if not *the* major goal, is to *solve* the system—that is, to use it to describe the evolution of the phenomenon. Solution of dynamical systems will therefore constitute another major emphasis of this book.

Building and solving dynamical systems are worthy ends in themselves, because of their powers of prediction. But dynamical systems also serve as engaging means of introducing, and developing, the fundamental constructs and concepts of calculus itself. In other words: the applications drive the mathematics just as much as the mathematics drives the applications. We think of this perspective as constituting a **contextual** view of calculus.

One thing about all of this that will, we hope, become clear later in this chapter is that *solution of dynamical systems can be computationally intensive*. In fact, it's completely impractical, in general, to try and solve these systems "by hand." For this reason we will, at many places in this book, apply computer mathematics software—we'll use the Sage package; packages like Mathematica and MATLAB work similarly—to the solution of these systems. We will discuss and develop the relevant computing techniques and notions as needed. (No previous computing experience is assumed.) Our approach to computing will be contextual too.

Certainly, not all of life, or even science, is about prediction. We often wish to study the past, or the present, as much as the future. Yet investigations of the past, and examinations of the present, are themselves often performed with an eye towards knowing and impacting the future. Moreover, analyses of the past—for example, approximating the age of an artifact using carbon dating—are often achieved through the application of predictive models that work just as well in either time direction.

1.2   The Spread of Disease: The SIR Model

Many human diseases are contagious: you "catch" them from someone who is already infected. Contagious diseases are of many kinds. Smallpox, polio, plague, and Ebola are severe and can be fatal, while the common cold and the childhood illnesses of measles, mumps, and rubella are usually relatively mild. Moreover, you can catch a cold over and over again, but you get measles only once. A disease like measles is said to "confer immunity" on someone who recovers from it.

Some diseases have the potential to affect large segments of a population; they are called *epidemics* (from the Greek words *epi*, upon + *demos*, the people.) *Epidemiology* is the scientific study of these diseases.

An epidemic is a complicated matter, but the dangers posed by contagion—and especially by the appearance of new and hard-to-control diseases, like COVID-19[1]—compel us to learn as much as we can about the nature of epidemics. Mathematics offers a very special kind of help.

First, we can try to draw out of the situation its essential features and describe them mathematically. Again, this process is called **(mathematical) modeling**. Second, we can use mathematical insights and methods to analyze the model. Any conclusion we reach about the model can then be interpreted to tell us something about the reality.

To give you an idea how this process works, we'll build a model—called the *SIR* model, for *susceptible, infected, recovered*—of an epidemic. This is a well-known, even iconic, model in epidemiology. Its basic purpose is to help us understand the way a contagious disease spreads through a population—to the point where we can even predict what fraction will fall ill, and when. Let's suppose the disease we want to model is like measles. Then it behaves (roughly) like the following.

1.2.1   Initial Setup

Our disease will entail the following three quantities, and their rates of change. These rates of change tell us "how fast we're going;" that is, they indicate the rates at which the indicated quantities are changing.

S: number of susceptible persons S': rate of change of S

I: number of infected persons I': rate of change of I

R: number of recovered persons R': rate of change of R

It's common in mathematics to denote the rate of change of a quantity Q by Q'. We'll have quite a bit to say about rates of change as we proceed. In fact, they will constitute a (the?) major focus of this book. For now, we'll be content to understand a rate of change intuitively, as the "speed" at which a quantity changes, or evolves. Note that a rate of change Q' will, itself, typically be changing, since Q might change quickly at some times and slowly at others, may be increasing at times and decreasing at others, and so on.

We make the following assumptions about the nature of this disease:

- Everyone recovers eventually.
- The duration of infection is the same for everyone.
- Once recovered, you're immune, and can no longer infect anyone.
- Only a fraction of contacts with the disease cause infection.
- The units of S, I, and R are persons.

---

[1] COVID-19 is generally described as a *pandemic*, meaning an epidemic that has spread across multiple countries or continents.

- The units of time are days.
- The units of S', I', and R' are persons per day, written persons/day.
- The system is *closed;* this simply means that the total size of the population, which equals the sum $S + I + R$, does not change.

1.2.2   Thinking About S', I', and R'

In this section, we will develop a system of *rate equations*—mathematical formulas describing rates of change—for S, I, and R. Later, we'll apply a technique known as *Euler's method* to this system, to predict the evolution of our epidemic.

Let's begin by addressing R'. We deal with this rate of change first because its analysis is, in many ways, the easiest of the three.

Suppose infection lasts for k days. Also assume, in the absence of any definite information to the contrary, that the infected population is "uniform with respect to duration of infection," at any given point in time. That is, assume that there are, any any instant, just as many people in this population who have been infected for one day as there are who have been infected for two, or three, and so on, up to k days.

Then on any given day, one kth of the infected population will recover. In other words, the *rate of recovery*, in persons per day, is equal to $1/k$ times I. Recall that we are calling this rate R'. So, in symbols:

$$R' = bI, \quad \text{where} \quad b = \frac{1}{k}.$$

Rate equation for the recovered population

Here, b is constant, in that it doesn't change over the course of time. Of course, different diseases may entail different values of b. In mathematical modeling, a number that is constant within a given situation, but may vary from situation to situation, is called a *parameter*. Often, we will use uppercase letters for the variables (other than the time variable) in rate equations, and will use lowercase letters for the parameters (and the time variable).

When one variable quantity equals a constant times another, we say the two quantities are *proportional*. So in our present model, the *rate of change* of the recovered population is proportional to the *size* of the infected population.

Let's move on to examination of S', which is the next easiest rate of change to model. Let's suppose that:

(i) Each susceptible person comes into contact with a proportion, call it p, of the infected population each day. This implies that each susceptible person has contact with pI infected persons per day. This in turn implies that there are $pI \cdot S = pSI$ *total* contacts between susceptible and infected persons each day.

(ii) A certain proportion, call it q, of the above contacts cause infection.

The above tells us that there are $q \cdot pSI$ new infections occurring each day. Which in turn implies that the size of the susceptible population *decreases* by $qpSI$ persons each day. In other words, in persons/day, we have

$$S' = -aSI, \quad \text{where} \quad a = qp.$$

Rate equation for the susceptible population

The minus sign denotes a *negative* rate of change, meaning a *decrease* in the quantity in question.

Finally, we consider I'. Since $S + I + R$ is assumed to be constant, the sum $S' + I' + R'$ of the rates of change of the three subpopulations must be *zero*—any change in one of these quantities is offset by changes in the others. That is, by the above formulas for R' and S',

$$0 = S' + I' + R' = -aSI + I' + bI$$

or, solving for I',

$$I' = aSI - bI.$$

Rate equation for infected population

To summarize: under the conditions described above, we have

$$\begin{aligned} S' &= -aSI \\ I' &= aSI - bI \\ R' &= bI \end{aligned}$$

The *SIR* equations

Here:

- The number a is a positive parameter called the *transmission coefficient*. Recall from above that $a = qp$, where p is the proportion, or fraction, of the infected population with which each infected person comes into contact each day, and q is the proportion these of contacts that cause infection. Note that the *units* of a are 1/(person-day), or "inverse person-days." Why? Because the units on both sides of any equation must agree. So, considering for example the equation $S' = -aSI$, we see that inverse person-days are indeed the correct units for a, because they insure that the right-hand side of this equation has units

$$\frac{1}{\text{person-day}} \cdot \text{persons} \cdot \text{persons} = \frac{\text{persons}}{\text{day}},$$

which are the units of the left-hand side of this equation.

- The number b is a positive parameter called the *recovery coefficient*. Recall from above that $b = 1/k$, where k is the number of days that infection lasts. The units of b are 1/day, or inverse days.

As noted above, S is a *decreasing* quantity. This is reflected in the above equation for S': the quantities a, R, and I are all positive, so $S' = -aSI$ is negative. And again, a negative rate of change corresponds to a decreasing quantity. Actually, it's conceivable that I or R might be zero at some point(s), in which case S' would be zero there too. So technically, it might be more precise to say that S is *nonincreasing*: its rate of change is never positive. To keep the terminology simple, though, we'll use the term "decreasing" even for quantities that are, strictly speaking, nonincreasing.

Similarly, R is an increasing quantity. The size I of the infected polulation can increase *or* decrease, since the sign of $I' = aSI - bI$ can be positive or negative, depending on the relative sizes of aSI and bI. In the next section, we'll look at the rise and fall of I more closely.

Also in the next section, we'll use our SIR equations, together with a generalized version of the *Euler's method* discussed in the previous section, to study the progress of our epidemic.

1.2.3   Exercises

1. The graph below depicts a disease evolving according to the above SIR model.

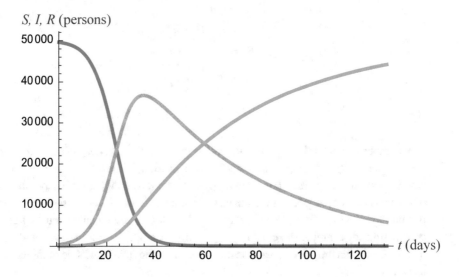

(a) Which curve is S, which is I, and which is R? Explain how you know which is which.

(b) About how long does it take before the susceptible population has decreased to half of its original size?

(c) About how large is the overall population (comprising all susceptible, infected, and recovered individuals)?

(d) About how long does it take before the infection peaks?

(e) About what is the size of the infected population at the time when infection peaks?

(f) About what is the size of the *susceptible* population at the time when infection peaks? (This value of S is called the *threshold value*, to be discussed in Sect. 1.4 below.)

(g) About how many people became infected over the course of the first 20 days? (Note: this is *not* the same as the number who are infected on day 20. Hint: look at the S curve.)

2. The graph below again depicts a disease evolving according to the SIR model.

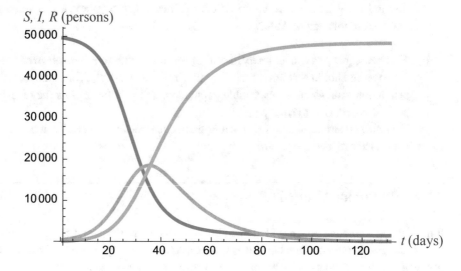

In this graph, the initial values $S(0)$, $I(0)$ and $R(0)$, and the transmission coefficient a, are the same as in the graph in Exercise 1 above. But the two graphs correspond to different recovery coefficients b.

Which of these two graphs corresponds to the *larger* value of b? Please explain.

3. A town of population 100,000 is hit with a measles epidemic, which evolves according to the above SIR equations. This unique strain of the measles is known to last for twelve days.

(a) What is the recovery coefficient b, and what are the units for b? Please explain.

Now suppose that, on day 15, 14,893 people are susceptible (that is, $S(15) = 14{,}893$) and 69,613 people are infected ($I(15) = 69{,}613$). Also suppose that, one tenth of a day later, the number of susceptibles has decreased to 13,856.

(b) What, at least approximately, is $S'(15)$ (the rate of change of S at $t = 15$)? What are the units of $S'(15)$? Hint: net change in a quantity (roughly) equals rate of change times elapsed time. (See Eq. (1.2) below.)

(c) What is the transmission coefficient a? (Hint: use information determined in part (b) of this exercise, together with the equation for S' appearing in the SIR model.) What are the units for a?

4. Consider an epidemic that progresses according to the usual SIR model, except that, now, recovered people become susceptible again (and can infect again) after m days. (The model for such a disease is sometimes called the $SIRS$ model, to reflect the fact that the recovered population feeds back into the susceptible population.)

Modify the usual SIR equations to reflect this new feature (wherein recovered can become susceptible again). Hints:

(i) Your new equations will look a *lot* like the old ones, but with some *new terms* added on. These terms should account for the facts that, now, on average, $1/m$ of the recovered population gets *added to* susceptible population, and *subtracted from* the recovered population, on any given day.

(ii) Your new equations should involve unspecified parameters a, b, and c, where a and b are as above, and $c = 1/m$.

1.3   Prediction Using SIR

Euler's method amounts to a **big idea** that allows us to use "rate equations," like the above SIR equations, to predict. That big idea is this: if Q is any quantity, varying with time, then between any two instants—a "new" one and an "old" one, say—we have

$$\text{New}\,Q = \text{Old}\,Q + \Delta Q \tag{1.1}$$

where ΔQ denotes the *change* in Q, from the "old" instant to the "new" instant. Moreover, we have

$$\Delta Q \approx Q' \Delta t \tag{1.2}$$

where Δt is the elapsed time, and Q' is the rate of change of Q with respect to time. Equation (1.2) simply says: **net change in a quantity (roughly) equals the rate of change of that quantity, times elapsed time.** And Eq. (1.1) simply says: **the new value of a quantity is the old value plus any change in going from old to new.**

Remark 1.3.1 The "$\approx$" in Eq. (1.2) above means "is approximately equal to." Why are the two sides of Eq. (1.2) only approximately equal to each other? Because Q' itself is (typically) changing. Distance traveled, for example, only equals velocity times time if the velocity is constant over the interval of time in question. One can think of things this way: if the velocity is changing, then reading the speedometer at a given instant will not allow us to predict the distance traveled over the next hour *exactly*—or even very well, most likely. On the other hand, it will probably give a pretty good idea of distance traveled over the next second, say.

The upshot is that Eq. (1.2) should be "pretty accurate" if Δt is small. How small, and how accurate? These questions will be explored as we proceed. For now, we'll understand Eq. (1.2) in the sense—admittedly, a somewhat vague sense—just discussed.

Remark 1.3.2 We pause a for moment to reflect on the (approximate) formula (1.2). In particular, we observe that this formula captures the above prediction principle—"If you know how fast you're going, then you know how far you'll get in a given amount of time"— mathematically, and quite succinctly. Indeed, (1.2) tells us: if you know Q' (that is, "how fast you're going"—remember that Q' is the rate of change of Q), then you know, at least approximately, what the change ΔQ in Q will be (that is, "how far you'll get") after Δt units have transpired (that is, "in a given amount of time").

For these reasons, we give Eq. (1.2) a name: we call it **the prediction equation**. This equation will be central to a large proportion of our studies.

1.3.1 An Example

Now, let's *use* Eqs. (1.1) and (1.2), together with the above SIR equations, to predict, as follows.

Example 1.3.1 Consider a disease that behaves according to the above SIR model. Suppose the initial values $S(0)$, $I(0)$, and $R(0)$ of S, I, and R, at time $t = 0$, are given by

$$S(0) = 500, \quad I(0) = 10, \quad R(0) = 0.$$

As before, we'll take the units of S, I, and R to be persons, and the units of time t to be days. Let's suppose we also know that the transmission and recovery coefficients a and b are given by

$$a = 0.001 \text{ (person-day)}^{-1}, \quad b = 0.2 \text{ day}^{-1}.$$

Use this information to predict $S(4)$, $I(4)$, and $R(4)$, using

(i) stepsize $\Delta t = 2$;
(ii) stepsize $\Delta t = 4$.

Solution.

(i) As we're starting at $t = 0$, and using stepsize $\Delta t = 2$, the first values of S, I and R to be predicted are $S(2)$, $I(2)$, and $R(2)$. Let's begin with $S(2)$. We have

$$
\begin{aligned}
S(2) &= S(0) + \Delta S && \text{(by (1.1))}\\
&\approx S(0) + S'(0)\Delta t && \text{(by (1.2))}\\
&= S(0) + (-aS(0)I(0))\Delta t && \text{(by the } SIR \text{ equations)}\\
&= 500 + (-0.001 \cdot 500 \cdot 10) \cdot 2 && \text{(plug in numerical values)}\\
&= 500 - 10 = 490.
\end{aligned}
$$

Note: Because an approximation occurs *somewhere* (anywhere!) in the computation of $S(2)$, the final result of that computation is itself an approximation. So, in spite of the "=" appearing in the last step (and in various other steps) of the above computation, what we have actually found is that $S(2) \approx 490$, and not that $S(2) = 490$.

Next, we compute $R(2)$ (we'll save $I(2)$ for last, because the equation for I' is less simple than the one for R'):

$$
\begin{aligned}
R(2) &= R(0) + \Delta R && \text{(by (1.1))}\\
&\approx R(0) + R'(0)\Delta t && \text{(by (1.2))}\\
&= R(0) + (bI(0))\Delta t && \text{(by the } SIR \text{ equations)}\\
&= 0 + (0.2 \cdot 10) \cdot 2 && \text{(plug in numerical values)}\\
&= 0 + 4 = 4.
\end{aligned}
$$

To find $I(2)$, we use the fact that, by assumption, $S + I + R$ is constant. Since, initially (at $t = 0$), this sum equals $500 + 10 + 0 = 510$, we have

$$
I(2) = 510 - S(2) - R(2) \approx 510 - 490 - 4 = 16.
$$

To summarize our first "step" of part (i) of this example: we've found that

$$
S(2) \approx 490, \quad I(2) \approx 16, \quad R(2) \approx 4 \quad \text{(persons)}. \tag{1.3}
$$

For the next step—estimating $S(4)$, $I(4)$, and $R(4)$—we imagine now that $t = 2$ is our "old," or starting, value of t, and that $t = 4$ is our "new," or final, value of t. We then proceed as above, using the (approximate) values of $S(2)$, $I(2)$, and $R(2)$ just computed, and summarized in Eq. (1.3). So, by the same kind of reasoning as we used in the first step,

$$
\begin{aligned}
S(4) &= S(2) + \Delta S\\
&\approx S(2) + S'(2)\Delta t\\
&= S(2) + (-aS(2)I(2))\Delta t\\
&\approx 490 + (-0.001 \cdot 490 \cdot 16) \cdot 2\\
&= 490 - 15.68 = 474.32.
\end{aligned}
$$

Note that, this time, we use the symbol "$\approx$" in two different instances—the first time because, as before, net change is only approximately equal to rate of change times elapsed time; the second time because the "old" values of S, I, and R that we're using (that is, the values at $t = 2$) are, themselves, approximations. (And never mind the 32 hundredths of a person who is presumably part of this susceptible population at $t = 4$. While math may be used to model real life, the two aren't the same, which is probably a good thing for that 0.32 of a person.)

Similarly,

$$\begin{aligned} R(4) &= R(2) + \Delta R \\ &\approx R(2) + R'(2)\Delta t \\ &= R(2) + (bI(2))\Delta t \\ &\approx 4 + (0.2 \cdot 16) \cdot 2 \\ &= 4 + 6.4 = 10.4, \end{aligned}$$

and

$$I(2) = 510 - S(4) - R(4) \approx 510 - 474.32 - 10.4 = 25.28.$$

In sum, then: using $\Delta t = 2$, we have found that

$$S(4) \approx 474.32, \quad I(4) \approx 25.28, \quad R(4) \approx 10.4 \quad \text{(persons)}. \tag{1.4}$$

(ii) In much the same manner as above, we find that

$$\begin{aligned} S(4) &= S(0) + \Delta S & R(4) &= R(0) + \Delta R \\ &\approx S(0) + S'(0)\Delta t & &\approx R(0) + R'(0)\Delta t \\ &= S(0) + (-aS(0)I(0)) \cdot \Delta t & &= R(0) + (bI(0))\Delta t \\ &\approx 500 + (-0.001 \cdot 500 \cdot 10) \cdot 4 & &\approx 0 + (0.2 \cdot 10) \cdot 4 \\ &= 500 - 20 = 480, & &= 0 + 8 = 8, \end{aligned}$$

and

$$I(4) = (S(0) + I(0) + R(0)) - S(4) - R(4) \approx (500 + 10 + 0) - 480 - 8 = 22.$$

In sum: using $\Delta t = 4$, we find that

$$S(4) \approx 480, \quad I(4) \approx 22, \quad R(4) \approx 8 \quad \text{(persons)}. \tag{1.5}$$

Compare (1.5) with (1.4): not surprisingly, these estimates are different. Again, the smaller stepsize $\Delta t = 2$ yields better results than $\Delta t = 4$, because the rates of change S, I, and R

are themselves continuously changing. And the smaller Δt is, the more often we recalibrate, to adjust for this change.

Using essentially the "Euler's method" algorithm implemented above, but with stepsize $\Delta t = 0.001$, we would find that

$$S(4) \approx 463.57, \quad I(4) \approx 31.30, \quad R(4) \approx 15.13 \quad \text{(persons)}.$$

These numbers are still approximations, but they are closer to the truth.

Note that, to approximate $S(4)$, $I(4)$, and $R(4)$ using stepsize $\Delta t = 0.001 = 1/1000$, we need to compute $S(t)$, $I(t)$, and $R(t)$ at $4 \cdot 1,000 = 4,000$ different values of t. Needless to say, we would not, and did not, do these computations by hand. We used a computer, together with the open-source Sage mathematical software package, which is very similar to MATLAB, Mathematica, and other mathematical software that you may have seen.

In Sect. 1.5, we'll discuss the use of Sage to these ends.

If we could somehow make sense of the above algorithm for the case $\Delta t = 0$, we would, in theory, have an *exact* solution to the SIR equations. As it turns out, the SIR system of rate equations does *not*, in fact, admit an exact, "closed-form" solution, meaning one where $S(t)$, $I(t)$, and $R(t)$ can be written as mathematical expressions in the variable t. Many other interesting "real-life" phenomena do. We'll discuss this further in the course of this text.

1.3.2   Summary: Euler's Method and SIR

Schematically, Euler's method, as applied to the SIR system of rate equations, looks like this (Fig. 1.1).

The starting point of this loop corresponds to using $S(0)$, $I(0)$, and $R(0)$ as the "current" values of S, I, and R. And what about the ending point? If one wants to predict all the way out to time T, say, and if one chooses a stepsize Δt, then one will need cycle through the above loop $T/\Delta t$ times. (Including the initial values $S(0)$, $I(0)$, and $R(0)$, one will then end up with $(T/\Delta t) + 1$ different values of S, I, and R.)

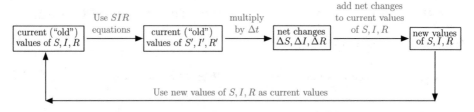

Fig. 1.1 Schematic diagram for Euler's method applied to the SIR equations

1.3.3 Exercises

1. We consider a measles epidemic with transmission coefficient $a = 0.000005$ (person-day)$^{-1}$, and recovery coefficient $b = 1/14$ day^{-1}. This epidemic is then modeled by the equations

$$S' = -0.000005\, SI,$$
$$I' = 0.000005\, SI - I/14,$$
$$R' = I/14.$$

We assume that the initial values of S, I, and R are:

$$S(0) = 45,400, \quad I(0) = 2,100, \quad R(0) = 2,500.$$

(a) Calculate the "current" rates of change $S'(0)$, $I'(0)$, and $R'(0)$, and use these rates of change to estimate $S(1)$, $I(1)$, and $R(1)$.

(b) Using the values of $S(1)$, $I(1)$, and $R(1)$ found in the previous exercise, calculate $S'(1)$, $I'(1)$, and $R'(1)$, and use these rates of change to estimate $S(2)$, $I(2)$, and $R(2)$.

(c) Using the values of $S(2)$, $I(2)$, and $R(2)$ found in the previous exercise, calculate $S'(2)$, $I'(2)$, and $R'(2)$, and use these rates of change to estimate $S(3)$, $I(3)$, and $R(3)$.

2. Go back to the starting time $t = 0$, and to the initial values and parameter values specified in Exercise 1 above. Recalculate the values of S, I, and R at time $t = 2$, this time using stepsize $\Delta t = 2$. You should perform only a single round of calculations, using the rates $S'(0)$, $I'(0)$, and $R'(0)$. Which estimates of $S(2)$, $I(2)$, and $R(2)$ do you think are "better:" those of this exercise, or those of Exercise 1(b) above?

3. Repeat Exercise 1 above, with the same initial values of S, I, and R, and the same transmission coefficient, but this time with recovery coefficient $b = 1/5$ day^{-1}.
How do your values of $S(3)$, $I(3)$, and $R(3)$ obtained here compare with those of Exercise 1(c) above? Explain these changes from a modeling perspective. That is: by considering the "real life" meaning of b, explain why it makes sense that changing b from $1/14$ to $1/5$ would result in the kinds of changes you see in $S(3)$, $I(3)$, and $R(3)$.

4. Repeat Exercise 1 above, with the same initial values of S, I, and R, and the same recovery coefficient, but this time with transmission coefficient $a = 0.00001$ (person-day)$^{-1}$.
How do your values of $S(3)$, $I(3)$, and $R(3)$ obtained here compare with those of Exercise 1(c) above? Explain these changes from a modeling perspective. That is: by considering the "real life" meaning of b, explain why it makes sense that changing a from 0.000005 to 0.00001 would result in the kinds of changes you see in $S(3)$, $I(3)$, and $R(3)$.

5. Repeat Exercise 1 above for a disease that evolves according to the $SIRS$ model of Exercise 4 of Sect. 1.2.3. Use the same initial values of S, I, and R, and the same transmission and recovery coefficients, as in Exercise 1 above, but this time, also assume

a time interval of 10 days between recovery and becoming susceptible again. (That is, in the language of Exercise 4 of Sect. 1.2.3, assume $m = 10$.)

How do your values of $S(3)$, $I(3)$, and $R(3)$ obtained here compare with those of Exercise 1(c) above? Explain these changes from a modeling perspective. That is: by considering the "real life" differences between the SIR and $SIRS$ models, explain why it makes sense that changing from one to the other would result in the kinds of changes you see in $S(3)$, $I(3)$, and $R(3)$.

1.4   More on the SIR Model

We next consider some basic implications of the above SIR model—implications that do not require Euler's method or any similar iterative process.

1.4.1 Threshold Value S_T of S

As noted earlier, the susceptible population only decreases in size. Eventually, this population will become small enough that it can no longer sustain growth in the infected population (assuming the latter population is, initially, growing). At this point, I will *peak*, and thereafter will dwindle.

The question that we wish to address is: how small is small enough? How small *does* the susceptible population need to become before I peaks, and begins to decline? We'll answer in a moment, but first, let's give a *name* to this particularly important value of S.

Definition 1.4.1 In the context of the above SIR model, the *threshold value of S*, denoted S_T, is the value of S at which I peaks.

Figure 1.2 gives a graphical interpretation of S_T. We can also use our above SIR equations—in particular, the equation for I'—to deduce a *formula* for S_T, as follows.

The condition "I peaks" in the above definition of S_T indicates that I changes from *increasing* to *decreasing*. In terms of rates of change, this means I' changes from *positive* to *negative*. Now it stands to reason that, at a point where a quantity changes from being positive to being negative, it must equal *zero*. (This reasoning fails if the quantity in question has some kind of sudden "jump" from a positive to a negative value, but let's assume this is not the case. In our above model, there's no reason to think I' would jump so abruptly.)

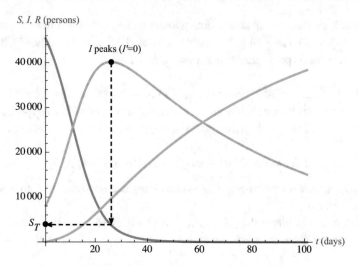

Fig. 1.2 Meaning of the threshold value S_T of S

So: S_T is the value of S where $I' = 0$. But by the above SIR equations, $I' = aSI - bI$. So we see that S_T satisfies the equation

$$aS_T I - bI = 0.$$

Factoring out the I gives

$$I(aS_T - b) = 0.$$

Assuming $I \neq 0$, so we can divide both sides of this equation by I to get

$$aS_T - b = 0.$$

Solving for S_T gives our final formula for the threshold value of S:

$$\boxed{S_T = \frac{b}{a}}$$

Threshold value S_T of S

For instance, in Example 1.3.1 above we have

$$S_T = \frac{0.2}{0.001} = 200.$$

In other words, as soon as the susceptible population has decreased from its initial value $S(0) = 500$ to just 200 remaining susceptible persons, the disease (specifically, the infected population) reaches its peak, and thereafter starts to wane.

Remark 1.4.1 At any point in time t where $S(t)$ is *smaller* than the threshold value S_T, I will be decreasing (as long as it is nonzero to begin with). To see this, suppose $S(t) < S_T$; that is, $S(t) < b/a$. Then, multiplying both sides by a, we get $aS(t) < b$, so $aS(t) - b < 0$, so

$$I'(t) = aS(t)I(t) - bI(t) = I(t)(aS(t) - b) < 0,$$

meaning $I'(t)$ is negative, and therefore I is decreasing at time t (and thereafter, since I can peak at most once).

Similarly, if $S(t)$ is *larger* than S_T, then I will be increasing at time t (but will start to decrease as soon as S becomes smaller than S_T).

1.4.2   Herd Immunity

One very nice application of the threshold value is to the understanding of *herd immunity*. This is the phenomenon where enough people have achieved immunity (through having been infected, through vaccination, etc.) that infection tapers off.

Herd immunity is often understood in terms of the following question: what *fraction*, or *proportion*, of the original susceptible population must become immune in order to ensure that the infected population decreases with time?

To answer note that, if a fraction f of the original $S(0)$ susceptible persons are immune, then the remaining number of susceptible people is $S(0) - fS(0)$. By Remark 1.4.1, infection will decrease if this number is less than $S_T = b/a$. This gives us an inequality that we can solve for f, as follows:

$$S(0) - fS(0) < \frac{b}{a}$$

$$S(0) < \frac{b}{a} + fS(0)$$

$$S(0) - \frac{b}{a} < fS(0)$$

$$\frac{S(0) - b/a}{S(0)} < f$$

$$1 - \frac{b}{aS(0)} < f.$$

To summarize:

Herd immunity occurs when the rate f

of immunity is larger than $1 - \dfrac{b}{aS(0)}$.

Criterion for herd immunity

For instance, the epidemic of Example 1.3.1 above satisfies

$$1 - \frac{b}{aS(0)} = 1 - \frac{0.2}{0.001 \cdot 500} = 0.6.$$

So we would require a rate of immunity greater than 60% to assure herd immunity.

1.4.3   Reproduction Number $r(t)$

We begin with the following.

Definition 1.4.2 The *reproduction number $r(t)$* is the number of new infections caused, on average, by each infected person, at time t.

In some references, the reproduction number is denoted $R(t)$ (or even R_t). We use a lower-case r to distinguish this number clearly from the variable R describing our recovered population.

The number $r(0)$, documenting new infections per infected person at the *outset* of the epidemic (at time $t = 0$), is often called the *basic reproduction number*.

We can derive a formula for $r(t)$ by way of the following observations. First: from the equation $S' = -aSI$, we see that the susceptible population at time t is *decreasing* at the rate of $aS(t)I(t)$ persons per day. So the number of new infections at time t is *increasing* at this same rate (since new infections are precisely the outputs from the susceptible population). But then the *rate* of new infections, measured in new infections *per infected person* per day, equals $aS(t)I(t)$ *divided by* the number of infected persons, which is to say it equals

$$\frac{aS(t)I(t)}{I(t)} = aS(t).$$

Again, this is the *daily rate* of new infections per infected person; to get the *total number* of new infections per infected person (on average), we multiply by the average duration of infection, which is k days, to get $kaS(t)$. To summarize,

$$\boxed{r(t) = kaS(t)}$$

Reproduction number $r(t)$

(Recall that $k = 1/b$, where b is the recovery coefficient.)

So, for instance, the *basic* reproduction number $r(0)$ for our epidemic of Example 1.3.1 is given by

$$r(0) = kaS(0) = (1/0.2) \cdot 0.001 \cdot 500 = 2.5.$$

In other words: at the outset of the epidemic, each infected person infects 2.5 others, on average.

Note that 2.5 is greater than one, so at least at the outset, each infected person is adding more to the infected population than will be lost through that person's eventual recovery. In such a situation, we would expect that I would continue to grow. On the other hand, if each infected person were to cause fewer than one new infection, we would expect I to taper off. That these expectations are, in fact, borne out is reflected in the following important result.

Proposition 1.4.1 *I is increasing at time t if $r(t) > 1$, and is decreasing at time t if $r(t) < 1$. That is, the infected population will grow whenever each infected person is infecting more than one other, and will drop whenever each infected person is infecting fewer than one other.*

The mathematical verification of the above proposition is outlined in Exercise 6 below. And in Exercise 7 below, we exhibit a simple relationship between herd immunity and the basic reproduction number.

1.4.4 Exercises

1.4.4.1 Part 1: Threshold Value; Increase/Decrease in Infection

1. **Quarantine**. For this exercise, assume that all conditions, parameters, and initial values are as described in Exercise 1 of Sect. 1.3.

   (a) What is the threshold value S_T for this epidemic?
       One of the ways to treat an epidemic is to keep the infected away from the susceptible; this is called quarantine. The intention is to reduce the chance that the illness will be transmitted to a susceptible person. Thus, quarantine alters the *transmission coefficient*.
   (b) Suppose a quarantine is put into effect that cuts in half the chance that a susceptible will come into contact with an infected. What is the new transmission coefficient? Explain how you know this. (Refer to the discussion of p and q beginning on Sect. 1.2.2.)

(c) Changing the transmission coefficient, as in part (b) of this exercise, changes the threshold level for S. What is the new threshold value S_T for this epidemic?

(d) Does this quarantine eliminate growth of the epidemic, in the sense that the number of infected immediately goes down from 2,100, without ever showing an increase in the number of cases? (Assume, again, that we start with we start with $S(0) = 45,400$.) Hint: see Remark 1.4.1 above.

2. **Other interventions**. Instead of (or in addition to) reducing the number of contacts between susceptible and infected individuals, one might want to reduce the likelihood that such a contact will lead to a new infection. Social distancing, and other measures like wearing face masks, can have such an effect.

   How will an intervention that cuts this likelihood of infection in half affect the transmission coefficient and the threshold value of the epidemic?

3. Construct the appropriate SIR model for a measles-like illness that lasts for 4 days. Assume it is also known that a typical susceptible person meets only about 0.3% of infected population each day, and the infection is transmitted in only one contact out of six. Hint: see the discussions of R' and S' in Sect. 1.2 above.

4. What is the threshold value for this epidemic?

5. How small does the susceptible population have to be for this illness to fade away from the outset? Hint: see Remark 1.4.1 above.

1.4.4.2 Part 2: The Reproduction Number and Herd Immunity

For the exercises in this part, the quantities S, I, R, k, b, a, and $r(t)$ are as in this section, above.

6. In this problem, we study the relationship between the basic reproduction number and growth of the infected population.

   (a) Use the equation for the basic reproduction number $r(t)$ on Sect. 1.4.3 to show that

   $$S(t) = \frac{b}{a} r(t).$$

   (b) Put the expression for $S(0)$ that you found in part (a) of this exercise into the formula for $I'(0)$ that you get from the SIR equations, and then do some algebra, to show that

   $$I'(0) = bI(0)(r(t) - 1).$$

   (c) Using part (b) of this exercise, explain how you know that Proposition 1.4.1 above is true. Hint: recall that $I' > 0$ tells us that I is increasing, while $I' < 0$ tells us that I is decreasing.

7. Suppose a fraction f of the initial susceptible population achieves immunity. Show that this will prevent infection from spreading as long as

$$f > 1 - \frac{1}{r(0)}.$$

Hint: Use the condition

$$f > 1 - \frac{b}{aS(0)}$$

given on Sect. 1.4.2 above, together with part (a) of Exercise 6 above.

8. Consider an SIR-like epidemic with susceptible population of initial size $S(0)$, and such that, at the outset, each infected person will infect four others, on average.

Suppose the susceptible population falls to 20% of its original size. Will this be enough to prevent infection from spreading—that is, to assure that I will decrease?

9. Consider the epidemic of Exercise 1 in Sect. 1.3 above.

(a) What is the basic reproduction number?
(b) How large a rate f of immunity would suffice to assure herd immunity at the outset? Express your answer in terms of percents, e.g. "Any rate greater than _____% would suffice."

1.5   Using a Program

1.5.1   Computers

A computer changes the way we can use calculus as a tool, and it vastly enlarges the range of questions that we can tackle. First, it is fast. It can do billions of additions in the time it takes us to do one. Second, it can be programmed. By arranging computations into a "loop"—a sequence of steps that may be performed repeatedly (see, for example, Fig. 1.1)—we can construct a program with only a few instructions that will carry out millions of repetitive calculations.

The purpose of this section is to give you practice using a computer program that estimates values of S, I, and R in the epidemic model. As you will see, it carries out the "Euler's method" calculations you have already done by hand. It also contains a loop that will allow you to do a many, many rounds of calculations with no extra effort.

1.5.1.1 The Program SIR

The program on the following page calculates values of S, I, and R that correspond to specified parameter values (a and b), starting and ending values of the time variable, and stepsize Δt. The program then plots these computed values of S, I, and R on a single graph.

The program is a set of instructions—sometimes called **code**—that is designed to be read by you and by a computer. These instructions mirror the operations we performed by hand, in the previous section, to generate *SIR* values.

Here and throughout this text, we present code written in the mathematical programming language know as Sage (or Sagemath). Sage is available for free use and has a web interface, and is therefore accessible and easy to use in the classroom.

Our Sage code here is quite similar to what it would look like in other mathematical programming languages, like Mathematica, MATLAB, and so on.

A computer reads the code one line at a time, starting at the top. Just about every line is a complete instruction that causes the computer to do something. The exceptions are the "comment lines," or lines starting with a hashtag ('#'). Comment lines are for human beings, rather than computers, to read. Comment lines *describe*, to the reader, what the adjacent code is intended to do. (When your computer "sees" a comment line, it ignores it.) Good commenting of computer code is essential for helping others make sense of the program (and for providing yourself with a reminder of what's going on). Read the comment lines below carefully to make sense of the program yourself!

The program SIR

```
# SIR program, for studying an epidemic using Euler’s method

# First, specify values of parameters, and initial values of variables

a=0.00001
b=1/14
S=49990
I=10
R=0

# Next, specify the starting and ending points in time, and the stepsize Deltat

t=0
tfin= 120
Deltat=0.5

# Set up lists for the values of S, I, R, and t that we’re about to compute. At first,
# these lists contain only the initial values of S, I, R, t specified above.

Svalues=[S]
Ivalues=[I]
Rvalues=[R]
tvalues=[t]

# The following loop does three things, repeatedly:

# (1) It computes the new values of S, I, R from the current values, using Euler’s method.
# (2) It increases t by the stepsize Deltat.
# (3) It stores the new values of S, I, R, and t into the lists created above.

while t<tfin:

    # Compute rates of change using SIR equations
```

```
Sprime=-a*S*I
Iprime=a*S*I-b*I
Rprime=b*I

# Net change equals rate of change times stepsize

DeltaS=Sprime*Deltat
DeltaI=Iprime*Deltat
DeltaR=Rprime*Deltat

# New values equal current values plus net change

S=S+DeltaS
I=I+DeltaI
R=R+DeltaR
t=t+Deltat

# Store new values

Svalues.append(S)
Ivalues.append(I)
Rvalues.append(R)
tvalues.append(t)

# Next time through the loop, the above new values
# play the role of current values
```

```
# Zip the t values with the S/I/R values into lists of ordered pairs,
# and create plots of these

Splot=list_plot(list(zip(tvalues, Svalues)), marker='.', size=50, color='blue',
legend_label="Susceptible", legend_color='blue')
Iplot=list_plot(list(zip(tvalues, Ivalues)), marker='.', size=50, color='red',
legend_label="Infected", legend_color='red')
Rplot=list_plot(list(zip(tvalues, Rvalues)), marker='.', size=50, color='green',
legend_label="Recovered", legend_color='green')

# Now plot the computed S, I, R values together on a single graph, with axes labelled
# appropriately

SIRgraph=Splot+Iplot+Rplot
show(SIRgraph, axes_labels=['$t$ (days)', '$S, I, R$ (persons)'])
```

We explain here some aspects of the above program. Further features—and some modifications—of the program are explored in the exercises below.

First: notice how, in the eleventh line (including comment lines but not including blank lines), Δt is typed out as `Deltat`. It is a common practice, in Sage (and many other languages), for the name of a variable to be several letters long.

The four lines starting with the line `Svalues=[S]` create lists into which we are going to store our successive values of S, I, R, and t. To begin with, each of these lists contains only one value: the initial value specified at the start of the program.

The line that says `while t<tfin:` plays a very important role. It says: repeat all of the *indented* lines that follow as long as the variable `t` has a value less than the specified ending value `tfin`.

The indentation is crucial here: all the indented lines are part of a "loop," meaning a sequence of lines to be repeated for a specified duration. Once the indentation stops, the computer "knows" not to interpret what follows as part of the loop. (You can use the "tab" key on your keyboard to indent.)

Now, consider the three indented lines starting with `Sprime=-a*S*I`. You should recognize them as coded versions of the rate equations

$$S' = -aSI$$
$$I' = \quad aSI - bI$$
$$R' = \qquad\quad bI$$

for the measles epidemic. (Sage uses $*$ to denote multiplication. Some languages do not. But Sage will not understand, for example, the input $7x$; you need to write $7*x$ instead.) These lines are instructions to assign numerical values to the symbols S', I', and R'. For example, the first of these three lines says:

Use the current values of S and I to compute the number $-aSI$; assign this number to S'.

The next two lines of code compute the current values of I' and R' similarly.

In other words, these three lines are performing the step indicated in the upper left corner of Fig. 1.1: they are using the SIR equations to compute current values of S', I', and R' from current values of S, I, and R.

Next is a comment line, followed by the three lines beginning with the line `DeltaS=Sprime*Deltat`. Together, these three lines do what the *top middle* step in Fig. 1.1 says to do: multiply each of the rates of change S', I', and R' by the stepsize Δt, to get the (approximate) *net changes* ΔS, ΔI, and ΔR.

There is then another comment line, followed by four lines that perhaps seem puzzling. Why would we write `S=S+DeltaS`, for example? Couldn't we cancel the t's and conclude `Deltat=0`?

The answer is no, because we should read these lines as computer instructions, rather than literal mathematics. As a computer instruction, `S= S+DeltaS` says

Use the current values of S and ΔS to compute the number $S + \Delta S$; let this number be the new value of S.

Once again we have an instruction that assigns a numerical value to a symbol, but this time the symbol (S, in this case) already has a value before the instruction is carried out. The instruction gives it a *new* value. (Here the value of S is changed by ΔS.)

In other words, these three lines of code that produce new values of S, I, and R correspond to the *top right* step in Fig. 1.1.

The indented line `Svalues.append(S)` says "take the current value of S, and put it at the end of the list that we've called `Svalues`." The idea is that, each time we compute a new value of S, we add it to this list. In this way, we'll end up, eventually, with a complete list (in chronological order) of our successively computed values of S.

The next three lines do similar things for the variables I, R, and t.

Again, because of the line `while t<tfin:`, new values of S, I, and R will be computed only as long as the *current* value of t is less than tfin. We don't need to compute new values when the current value of t is *equal to* tfin, because these new values would actually correspond to a t value just beyond the specified ending value `tfin`.

Words like "new" and "current" aren't needed in the computer code because they are automatically understood to be there. A computer instruction of the form A = B is always understood to mean "new A = current B."

Notice that the instructions A = B and B = A mean different things. The second says "new B = current A." Thus, in A = B, A is altered to equal B, while in B = A, B is altered to equal A.

Finally, we come to the lines of code *following* the loop. Together, these lines produce a graph comprising an S curve in blue, an I curve in red, and an R curve in green. The generated graph also contains a legend and labelled axes. We make just a few specific comments about these various plotting commands:

- The command `list_plot` is used to plot a list of ordered pairs (x, y), with the x values on the horizontal axis and the y values on the vertical axis.
- In our program, we have created lists of values of S, I, R, and t; we have *not* created lists of ordered pairs. The `zip` command helps us out here. For example, the command `zip(tvalues, Svalues)` pairs each element of the list `tvalues` with the corresponding element of the list `Svalues`. (By "corresponding" we mean "at the same location along the list.") Warning: problems will arise if you try to zip together two lists of different lengths.
- Note the curious dollar signs in the last line of the program. These have the effect of *italicizing* the variable names that they enclose. Technically, the dollar signs are invoking a computer typesetting language called LATEX (which is the language used to typeset this book). It's traditional, when typesetting mathematics, to italicize variable names.

You can better understand the various parameters in the above `plot` commands by modifying them, and observing the effects. In general, one of the the best ways to learn the finer points of coding is through experimentation.

1.5.2   Exercises

1.5.2.1 Part 1: SIR Using Euler's Method and Sage

For these exercises, you will study an SIR epidemic using Euler's method on a computer. You will do so by running, thinking about, and modifying the Sage program SIR, the code for which appears directly above.

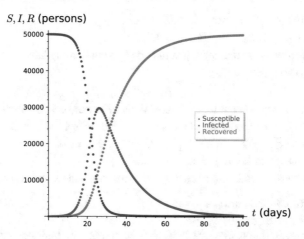

1. This exercise is meant simply to make sure your SIR program is working properly, and to get you thinking about coding in Sage.

 (a) Run your SIR program. Your graphical output should look something like the graph above.

   For the rest of these exercises let's assume that, as in the above graph, our independent time variable t is measured in days, and that our dependent variables are measured in numbers of persons.

 (b) In this model, what are the values of the transmission and recovery coefficients, and what are the initial values of S, I, and R? On average, how long does a person remain infected? What is the threshold value S_T of S? Use the program to answer.

 (c) What are the beginning and ending values of t? What is the stepsize Δt? At how many total points in time will we record observations? Use the program to answer. (You may want to recall the second paragraph of Sect. 1.3.2.) Your answers should be *numbers*, like "17," not variable names like "tfin."

 (d) Write down the first four values of t (including $t = 0$) at which observations will be made and recorded. Also write down the last four.

 (e) What are the indented lines of your program doing? Describe what computations are being done, and how many times these computations are being executed. You may want to think about the diagram in Fig. 1.1.

Estimate how long it would take you to do all of these calculations by hand (using a calculator that can only do $+$, $-$, $\times$, and $\div$). A VERY ROUGH ballpark estimate is fine, but do describe how you came up with that estimate.

2. After you have generated the above graph, go back to your program, and type in the following line of code, **right after** the existing line that says `tvalues.append(t)`:

```
# Print new values of t and S
print(t, S)
```

These lines need to be indented, just like the lines that immediately precede and follow it. Now run your program again.

(a) Describe the new output that you see. What information is being displayed?

(b) You may have received a message to the effect of "`WARNING: Output truncated!`" when you ran part (a) of this exercise. This is because you asked Sage to show a rather long list.

Now go back to the line `print(t, S)` that you added, and replace it with this:

```
if t>=20 and t<=25:
  print(t, S)
```

Again, indentation is important. Your code should now look like this:

```
tvalues.append(t)

# Print new values of t and S

if t>=20 and t<=25:
    print(t, S)

# Next time through the loop, the above new values
# play the role of current values

# Zip the t values with the S/I/R values into lists of
# ordered pairs, and create plots of these
```

Now run your program again. What do you see? What is the range of t values that are being displayed? (It's possible that your output does not include the data corresponding to $t = 20$, or to $t = 25$, or both, even though you specified that you wanted values between 20 and 25 inclusive. This might be due to the way the program treats decimal numbers. That is, your program might be storing the t-value $t = 25$ as 25.0000000000002, for example, and this would not be included in the range $20 \le t \le 25$. If this is occurring, can you think how to fix it?)

(c) Replace the statement if t > =20 and t < =25: from part (b) of this exercise with the line

```
if t > =30-Delta/2 and t < =30+Deltat/2:
```

and run your program again. Describe the numerical output that you get. Why do you think we built a "margin of error" of size $\Delta t/2$ on either side of the target t value of $t = 30$? (Hint: see the discussion at the end of part (b) of this exercise, above.)

3. Now, in the empty input box that appears at the bottom of your program, enter and run the code

```
SIRgraph.save('SIR-plot.pdf')
```

What did this line of code do for you?

4. Run your SIR program again, but this time, with new stepsize $\Delta t = 0.05$ instead of the original stepsize found in the code (and all other quantities the same as above).

(You may delete the new code that you added in Exercise 2 above. However, you may to need similar lines of code to answer some of the questions below that refer to numerical values of S, I, R, and/or t.)

How is the output you got in this case different from that of Exercise 1 above (besides the fact that the dots are more closely spaced in your new graph)? (It may help to look closely at, among other things, where I peaks in each of your two figures. You could also compare numerical values, using the ideas of Exercise 2 above.)

Explain why the two graphs should be different. Which of the two graphs do you think is "better," in the sense of giving a closer approximation to reality? Why?

For the remaining exercises in this part, use a stepsize of 0.05.

5. Run your SIR program again, but this time, with $b = 1/28$ instead of $b = 1/14$ (and all other quantities the same as in Exercise 2 above).

(a) What are the changes in the graphs of S, I, and R, relative to the graphs in Exercise 4 above? Describe in general terms; you don't have to discuss specific numerical values, although you can if you want.

(b) From a modeling perspective (that is, in terms of the "real life" interpretation), what's the meaning of the recovery coefficient $b = 1/28$? Explain, from a modeling perspective, why it makes sense that changing b from $1/14$ to $1/28$ would cause changes like the ones you saw in the graphs of S, I, and R.

(c) What is your new threshold value S_T? How does this compare to your answer from Exercise 1(c)? That is, which of these threshold values is larger, and by how much? Explain why this makes sense, from a modeling/"real-world" perspective.

6. **Quarantine and flattening the curve.** Reset b to the value $b = 1/14$.

The effect of quarantine is to decrease the likelihood of contact of a susceptible person with an infected person.

(a) Suppose a quarantine is imposed that cuts this likelihood in half. Which of the two parameters, a or b, will change as a result of this, and by how much will it change? Hint: you may want to refer to the discussions of p, q, and a in Sect. 1.2.2 above. Once you have figured out the answer, make the corresponding change to your SIR code, and run the program again to generate a new graph.

(b) Describe how this change affects your I curve, relative to the earlier graph of I in Exercise 4 above. In particular, what happens to the peak of the I curve? Does this peak happen sooner, or later, than it did before? Is this peak higher, or lower, than it was before?

(c) Describe how this change affects *the total number of people who get infected over the course of the disease*. Hint: don't look at the I curve for this, because a value of I reflects the numbers of people infected on a given day, and its hard to deduce from this how many become infected in total. Instead, look at the S curve. The number of susceptibles at the outset, minus the number of susceptibles at the end, tells you how many became infected over the course of the disease. (In your most recent code, reflecting the effects of quarantine, you might need to graph over a longer interval of t values, to get a good picture of how S is stabilizing.)

(d) What does quarantine seem to affect more dramatically: the number of people who ultimately become ill, how long it takes before the illness peaks, or the maximum number of people who can become ill at the same time? (Your answer may include more than one of these three phenomena.)

7. Reset all parameters to the values of Exercise 4 above. (So $\Delta t = 0.05$, $a = 0.00001$, $b = 1/14$.) We are now going to modify the SIR model so that recovered become *susceptible again* after 10 days. This is sometimes called the $SIRS$ model—the idea is that, in this case, immunity doesn't last forever, so the "R" population feeds back into the "S" population. (See Exercise 4 of Sect. 1.2.3.)
To do this:

(a) Make the appropriate changes to the program, to reflect this new phenomenon where recovered become susceptible again. You can do this by adding a line of the form "c=" after the line "b=" near the top of your program, and then by changing just *two more lines* of code in your program. (To be thorough, you should also modify some of your comment lines to reflect the new scenario.) Then execute the new code.

(b) Explain what changes you made to your SIR code to get your new program. You can do this by describing these changes in a brief paragraph, or by just specifying which lines you changed, and writing down what you changed them to.

(c) What are the changes in the graphs of S, I, and R, relative to the graphs in Exercise 4 above? Describe in general terms; you don't have to discuss specific numerical values, although you can if you want. Explain, from a modeling perspective, why it makes sense that the changes you made to the code would cause changes like the ones you saw in the graphs of S, I, and R.

8. Notice that, in the graph you generated in Exercise 7 above, values of I level off at a higher level than values of R. What *single parameter* would you change, and to what would you change it, to make I and R level off at the *same* height? (You might change a certain parameter to make I level off at the height of R; or you might change a certain parameter to make R level off at the height of I. Either way is fine.) Explain why this makes sense from a modeling perspective.

Once you've figured out how to answer the above question, make the required changes to your program, and run it.

1.5.2.2 Part 2: Short Programs to Practice on

9. Here are three short Sage programs for you to analyze and run. Please try to answer the questions about each program's output *before* running the program. Of course, you may then run the program to verify your answers.

Program 1	Program 2	Program 3

```
A=2              A=2              A=2
B=3              B=3              B=3
A=B              B=A              A=A + B
print(A, B)      print(A, B)      B=A + B
                                  print(A, B)
```

(a) When Program 1 runs, it will print the values of A and B that are current when the program stops. What values will it print? Type in this program and run it to verify your answers.

(b) Repeat part (a) of this exercise for Program 2.

(c) Repeat part (a) of this exercise for Program 3.

10. The next three programs use "`while`" statements, as in the above SIR program. Predict the output of each of these programs *before* running them. Then run them to confirm your answer.

What are the key differences between the three programs that cause the differences in output?

Program 4	Program 5	Program 6

```
k=0                k=0                k=0
while k <=5:       while k <=5:       while k <=5:
    A=k^2              k=k+1              A=k^2
    print(A)           A=k^2              k=k+1
    k=k+1              print(A)           print(A)
```

11. The next three programs have an element not found in the program SIR. They use "for" statements instead of "while" statements to designate a loop.

Run Program 7 and examine the output. Using this information, try to predict the output of each of programs 8 and 9 *before* running them. Then run them to confirm your answer. What are the key differences between the three programs that cause the differences in output?

Program 7	Program 8	Program 9

```
for k in [1..5]:   for k in [1..5]:   for k in [2..5]:
    A=k^3              A=k^3              A=k^3
    print(A)           print(A)           print(A)
```

1.6   Functions

A number of important mathematical ideas have already emerged in our study of an epidemic. In this section we pause to consider them, because they have a "universal" character. Our aim is to get a fuller understanding of what we have done so far, so we can use the ideas in other contexts.

One of these crucial ideas, which is central to mathematics, is that of a *function*. This idea is worth highlighting:

> **A function is a rule that specifies how**
> **the value of one variable, the input, determines**
> **the value of a second variable, the output.**

Definition of a function

That is, a function describes how one quantity depends on another. For example, in our study of a measles epidemic, the relation between the number of susceptibles S and the time t is a function. We write $S(t)$ to denote that S is a function of t. Here, the variable t is called the **input**, and the variable S is called the **output**. We think of S as *depending on t*, so t is also called the *independent variable* and S the *dependent variable*.

We also write $I(t)$ and $R(t)$, because I and R are functions of t, too. And we write $S'(t)$ to indicate that the rate S' at which S changes over time is a function of t.

Notice we say that a function is a *rule*, and not a *formula*. This is deliberate. We want the study of functions to be as broad as possible, to include various ways in which one quantity can be related to another.

So far, we have followed the standard practice in science of letting the single letter S designate both the *function*—that is, the *rule*—and the *output* of that function—that is, the *dependent variable*. Sometimes, though, we will want to make the distinction. In that case we will use two different symbols. For instance, we might write $S = f(t)$; here, we are still using S to denote the output, but the new symbol f stands for the function rule. Or we might write $y = S(t)$, in which case we are still using S to denote the rule, but the new symbol y stands for the output.

Example 1.6.1 Some other examples of functions are as follows:

1. The amount of postage you pay for a letter is a function of the weight of the letter.
2. The time of sunrise is a function of what day of the year it is.
3. The volume of a cubical box is a function of the length of a side. The last is a rather special kind of function because it can be described by an algebraic formula: if V is the volume of the box and s is the length of a side, then $V(s) = s^3$.
4. The formula $y = x^2$ defines y as a function of x. We have not given an explicit name to this function, but of course we could: we might call the rule f, in which case we could write $f(x) = x^2$ or, to be even more complete, $y = f(x) = x^2$. Or we might avoid introducing a new letter, like f, and simply use the same letter y to denote the both the output of the function and the function itself. That is, we might sometimes write $y(x) = x^2$.
   Similarly, the formulas $y = \sqrt{x-1}$, $y = 1/\sqrt{3-2x}$, and $y = 3x - 5$ all express an output y as a function of an input x. The last of these formulas gives an example of a *linear* function. Linear functions will be discussed in detail in the next section.
5. Temperature F, in degrees Fahrenheit, is a function of temperature C, in degrees Celsius, according to the formula
$$F = \frac{9}{5}C + 32$$
   (which also describes a linear function).
6. The formula
$$P(t) = \frac{100}{1 + 9e^{-t/10}}$$
might express population $P(t)$, in thousands, as a function of time t, in days (from a given starting point), in a certain "logistic growth" situation. We'll discuss the "exponential" function e^x that appears here, as well as logistic growth, in more detail later.

7. A **constant function** is one that gives the same output for every input. If h is the constant function that always gives back 24, then in formula form we would express this as $h(x) = 24$. Here, it doesn't matter what the input x is. For example, $h(0) = 24$, $h(-35) = 24$, $h(47\pi) = 24$, $h(\text{whatever}) = 24$.

8. Water density D, in kilograms per cubic meter (kg/m^3), is a function of water temperature C, in degrees Celsius; we might write $D = q(C)$.

1.6.1 Some Technical Details

Domain and range. The set of values that the input to a function takes is called the **domain** of the function. The domain may depend on the contexts, both physical and mathematical. If no physical context is given, then the domain is sometimes called the **natural domain**. This terminology is perhaps a bit misleading, in that the natural domain is the domain that applies in the *absence* of any natural, "real-world" constraints. But it is what it is.

For example, the function defined by $y = 1/\sqrt{3 - 2x}$ has natural domain equal to

$$\{\text{real numbers } x : x < 3/2\} \tag{1.6}$$

(the set of all real numbers x that are less than 3/2). Why? Because, mathematically, one can neither divide by zero nor take the square root of a negative number. So in the formula $y = 1/\sqrt{3 - 2x}$, we can neither have $3 - 2x = 0$ or $3 - 2x < 0$. So we must have $3 - 2x > 0$, or $3 > 2x$, or $3/2 > x$, or $x < 3/2$. In interval notation, the set of such x may be denoted $(-\infty, 3/2)$.

Next, consider the formula $F = \frac{9}{5}C + 32$ in item 6 of the above example. All by itself, this formula defines a function with natural domain equal to the set of all real numbers, denoted $(-\infty, \infty)$ or, sometimes, $\mathbb{R}$. This is because $\frac{9}{5}C + 32$ makes mathematical sense for any real number C.

But in this case, the formula is not the complete picture. A physical context for the formula $F = \frac{9}{5}C + 32$ was explicitly stated, which puts restrictions on our domain. Namely, since $-273\,^\circ$C is absolute zero, a reasonable domain to ascribe to this situation is

$$(-273, \infty) \quad \text{or} \quad \{C \in \mathbb{R} : C > -273\}.$$

One might argue that absolute zero is theoretically attainable, in which case one might take the domain to be $[-273, \infty)$. At the other end, contemporary models postulate a maximum attainable "Planck temperature" T_P equal to about $1.417 \cdot 10^{32}\,^\circ$C, so maybe the domain here should be $(-273, 1.417 \cdot 10^{32})$. (Or $[-273, 1.417 \cdot 10^{32}]$?)

The moral is that domains can sometimes be open to interpretation! (This is true even of *natural* domains. For example, if one allows for *imaginary* or *complex numbers*, then one *can* take the square root of a negative number. In this text, though, we will allow only for real numbers.)

The set of values taken by the output of a function is called the **range** of the function. This will depend on the domain. For example, if we take $[-273, 1.417 \cdot 10^{32}]$ as the domain of a function given by the formula $F = \frac{9}{5}C + 32$, then the range of this function is

$$\left[\frac{9}{5}(-273) + 32, \ \frac{9}{5}(1.417 \cdot 10^{32}) + 32 \right] = [-459.4, \ 2.5506 \cdot 10^{32}].$$

(The range of a function f, with domain $[a, b]$, is not always equal to $[f(a), f(b)]$. For example, the function $f(x) = x^2$, with domain $[-2, 2]$, does not have range $[f(-2), f(2)] = [4, 4]$. Rather, it has range $[0, 4]$. It's not always obvious how to get the range from the domain.)

Input versus output. Note the use of the words "rule," "specifies," and "determines" in our definition of function, above. These words all highlight an *essential* property of any function: a function associates a *unique* output to each particular input. Another word for "unique," in this context, is "unambiguous."

For example, the formula $y = x^2$ defines y as a function of x, because given x, we know exactly what y is: it's the square of x. If $x = 3$, we know unambiguously that $y = 3^2 = 9$, and so on.

As a consequence of our definition of function, the statement "y is a function of x" **need not** imply that x is a function of y. Indeed, the formula $y = x^2$ does not give x as a function of y. If we choose $y = 16$, for example, there is ambiguity as to what x must be: x *could* equal 4, but it could also equal -4, since both of these numbers satisfy the equation $16 = x^2$.

Of course, many functions *do* specify input and output uniquely in terms of each other. Such functions are sometimes said to be *one-to-one*. For example, the equation $F = \frac{9}{5}C + 32$ describes a one-to-one function; this function does give C uniquely in terms of F (in addition to giving F uniquely in terms of C). Specifically, we can solve this equation for C to get the unambiguous formula

$$C = \frac{5}{9}(F - 32).$$

We'll return to the topic of one-to-one functions in a later chapter.

1.6.2   Function Notation; Chaining, or Composing, Functions

It is important not to confuse an expression like $S(t)$ with a product; $S(t)$ does *not* mean $S \times t$. On the contrary, the expression $S(1.4)$, for example, stands for the output of the function S when 1.4 is the input. In the epidemic model, we interpret this as the number of susceptibles that remain 1.4 days after today (or whatever day we designate as $t = 0$).

The symbols we use to denote the input and the output of a function are just names; if we change them, we don't change the function. For example, here are four ways to describe the same function g:

$$g : \text{multiply the input by 5, then subtract 3};$$
$$g(x) = 5x - 3;$$
$$g(u) = 5u - 3;$$
$$g(\text{whatever}) = 5 \cdot \text{whatever} - 3.$$

It is important to realize that the *formulas* we just wrote in the last three lines are merely shorthand for the instructions stated in the first line.

If you keep this in mind, then complex-looking combinations like $g(g(2))$ can be decoded easily by remembering g of *anything* is just 5 times that anything, minus 3. We could thus evaluate $g(g(2))$ from the inside out:

$$g(g(2)) = g(5 \cdot 2 - 3) = g(7) = 5 \cdot 7 - 3 = 32,$$

or we could evaluate it from the outside in:

$$g(g(2)) = 5g(2) - 3 = 5(5 \cdot 2 - 3) - 3 = 5 \cdot 7 - 3 = 32.$$

Suppose f is some other rule, say $f(x) = x^2 - 1$. Remember that this is just shorthand for "Take the input (whatever it is), square it, and subtract 1," or "f of whatever is whatever squared, minus one." We could then evaluate

$$f(g(3)) = f(5 \cdot 3 - 3) = f(12) = 12^2 - 1 = 144 - 1 = 143,$$

while

$$g(f(3)) = g(3^2 - 1) = g(8) = 5 \cdot 8 - 3 = 37.$$

More generally, we have

$$f(g(x)) = f(5x - 3) = (5x - 3)^2 - 1 = 25x^2 - 30x + 9 - 1 = 25x^2 - 30x + 8$$

and

$$g(f(x)) = g(x^2 - 1) = 5(x^2 - 1) - 3 = 5x^2 - 5 - 3 = 5x^2 - 8.$$

When we use the output of one function as the input for another, we say that we are *chaining*, or *composing*, these functions. More specifically, the function $y = f(g(x))$ obtained by taking the output of g, and using this output as the input to f, is called *the composition* of f and g (denoted $f \circ g$ in some references, though we will not use this notation). Schematically, the picture is given in Fig. 1.3.

Warning: the composition of f and g is not, in general, the same as the composition of g and f; our above example illustrates a situation where $f(g(x)) \neq g(f(x))$.

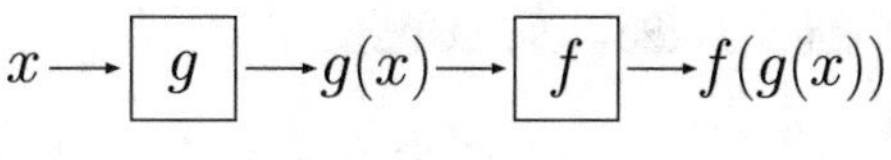

Fig. 1.3 The composition, or chain, of f and g

$$x \longrightarrow \boxed{g} \longrightarrow g(x) \longrightarrow \boxed{f} \longrightarrow f(g(x))$$

To avoid possible confusion, we will usually say things like "the composition $f(g(x))$" rather than "the composition of f and g," since the latter terminology does not emphasize the order is which the composition is done.

Note that chaining is not limited to situations involving only two functions. We can chain together any number of functions: For example, if f and g are as above, and $h(x) = 3 - x$, then we can form the function $y = g(h(f(x)))$, defined as follows (working from the inside out):

$$g(h(f(x))) = g(h(x^2 - 1)) = g(3 - (x^2 - 1)) = g(3 - x^2 + 1)$$
$$= g(4 - x^2) = 5(4 - x^2) - 3 = 20 - 5x^2 - 3 = 17 - 5x^2.$$

And so on.

Chaining will turn out to be very important later in this course. For now, though, you should treat it simply as part of the formal language of mathematics. It is somewhat analogous to conjugating verbs in a Japanese language class—it's perhaps not very exciting for its own sake, but it allows us to read the interesting stuff later on.

1.6.3 Functions of Several Variables

Many functions depend on more than one variable. For example, sunrise depends on the day of the year but it also depends on the latitude (position north or south of the equator) of the observer. (So when we said, in Example 1.6.1, that "The time of sunrise is a function of what day of the year it is," we perhaps should have added "all other things being equal.") Likewise, air temperature depends on both time and place. The postage necessary to mail a package depends on weight, dimensions, and other factors like the delivery method. And so on.

Functions of several variables can be given by formulas too. For example, the function

$$f(x, y) = 5x - y^2$$

takes two different real number inputs x and y, and outputs $5x - y^2$. One calculates values of such functions in a straightforward way, e.g.

$$f(7, -3) = 5 \cdot 7 - (-3)^2 = 35 - 9 = 26.$$

Note that the order matters:

$$f(-3, 7) = 5 \cdot (-3) - 7^2 = -15 - 49 = -64 \neq f(7, -3).$$

Similarly, one can define functions of three or more variables. And many of the processes that we've considered for functions of a single variable carry over to the multivariable situation. For example, if $g(x) = 3x + 2$, then we can chain, or compose, g with f, as follows:

$$g(f(x, y)) = g(5x - y^2) = 3(5x - y^2) + 2 = 15x - 3y^2 + 2.$$

On the other hand, note that we can't compose in the other order: whatever we input to g, the output will be a single real number, which will not suffice as input to f, because f requires two real inputs.

However, if we also define $h(y) = y^2$, then we can compose f with the pair (g, h):

$$f(g(x), h(y)) = f(3x + 2, y^2) = 5(3x + 2) - (y^2)^2 = 15x + 10 - y^4.$$

1.6.4   Exercises

1.6.4.1 Part 1: Functions: Evaluation, Composition, Domain

The next four exercises refer to these functions:

$$c(x, y) = 17 \qquad\qquad\qquad \text{a constant function}$$

$$j(z) = z \qquad\qquad\qquad \text{the identity function}$$

$$r(u) = 1/u \qquad\qquad\qquad \text{the reciprocal function}$$

$$P(a, b) = ab \qquad\qquad\qquad \text{the product function}$$

$$D(a, b) = a - b \qquad\qquad\qquad \text{the difference function}$$

$$s(y) = y^2 \qquad\qquad\qquad \text{the squaring function}$$

$$v(y) = \sqrt{y} \qquad\qquad\qquad \text{the square root function}$$

$$\ell(x) = 3 - x \qquad\qquad\qquad \text{a linear function}$$

$$Q(v) = \frac{2v + 1}{3v - 6} \qquad\qquad\qquad \text{a rational function}$$

$$H(x) = \begin{cases} 5 & \text{if } x < 0 \\ x^2 + 2 & \text{if } 0 \le x < 6 \\ 29 - x & \text{if } 6 \le x \end{cases} \qquad\qquad \text{a ``piecewise'' function}$$

$$T(x, y) = r(x) + Q(y)$$

1. Determine the following values:

$$c(5, -3) \quad s(17) \quad c(a, b) \quad j(u^2 + 1) \quad r(s(-4)) \qquad\qquad r(Q(3))$$
$$j(c(3, -5)) \quad \ell(1.1) \quad r(1/17) \quad Q(0) \qquad\quad r(r(v(r(r(u))))) \quad Q(r(3))$$
$$Q(2) \qquad\quad Q(3/7) \quad D(5, -3) \quad D(-3, 5) \qquad \ell(\text{whatever}) \qquad T(3, 7)$$
$$H(1) \qquad\quad H(7) \quad \ell(v(4)) \quad H(H(H(-3))) \quad D(\text{mellow, yellow}) \quad T(s(2), j(3))$$

2. True or false. Give reasons for your answers: if you say true, explain why; if you say false, give an example that shows why it is false.

   (a) For every non-zero number x, $r(r(x)) = j(x)$.
   (b) For every real number x, $v(s(x)) = x$.
   (c) For every positive number t, $s(v(t)) = t$.
   (d) $c(\pi, -965.32) = j(17)$.
   (e) If $a > 1$, then $s(a) > 1$.
   (f) If $a > b$, then $s(a) > s(b)$.
   (g) For all real numbers a and b, $s(a + b) = s(a) + s(b)$.
   (h) For all real numbers x, $s(r(x)) = r(s(x))$.
   (i) For all real numbers x, $s(\ell(x)) = \ell(s(x))$.
   (j) For all real numbers a, b, and c, $P(P(a, b), c)) = P(a, P(b, c))$.
   (k) For all real numbers a, b, and c, $D(D(a, b), c)) = D(a, D(b, c))$.

3. Find all numbers x for which $Q(x) = r(Q(x))$.
4. Recall that the **natural domain** of a function f is the largest possible set of real numbers x for which $f(x)$ is defined. For example, the natural domain of $r(x) = 1/x$ is the set of all *non-zero* real numbers.

   (a) Find the natural domains of $Q(x)$ and $H(x)$.
   (b) Find the natural domains of $P(z) = Q(\ell(z))$ and $R(v) = \ell(Q(v))$.
   (c) Find the natural domain of $W(y) = Q(v(\ell(y)))$.
   (d) What is the natural domain of the function $W(t) = \sqrt{\dfrac{1 - t^2}{t^2 - 4}}$?
   (e) Find functions $f(x)$ and $g(x)$ such that

   $$f(g(x)) = \sqrt{3 + (x + 2)^5} \quad and \quad g(f(x)) = (\sqrt{x + 3} + 2)^5.$$

1.6.4.2 Part 2: Graphing Functions Using Sage
Exercises 5–13 are intended to give you some experience using Sage for graphing. Using the Sage "plot" command, you can draw the graph of a function $y = f(x)$ whose formula you

know. You must type in the formula, using the following symbols: $+$, $-$, $*$, $/$, and $\char`^$ to represent addition, subtraction, multiplication, division, and exponentiation respectively. Here is an example:

to enter:	type:
$\dfrac{9x^5 - 5x^3}{x^2 + 1}$	`(9*x^5-5*x^3)/(x^2+1)`

(Recall that, in Sage you need to use $*$ to indicate multiplication.) The parentheses here are important. If you do not include them, the computer will interpret your entry as

$$9x^5 - \frac{5x^3}{x^2} + 1 = 9x^5 - 5x + 1 \neq \frac{9x^5 - 5x^3}{x^2 + 1}.$$

To graph the above function, with axes labeled, you can enter the code

```
plot((9*x^5-5*x^3)/(x^2+1), axes_labels=['$x$', '$y$'])
```

Sage will choose for you an interval of x values for the graph. If you want specify that the function should be graphed over the interval $-2 \le x \le 2$, you can input the code

```
plot((9*x^5-5*x^3)/(x^2+1), -2, 2, axes_labels=['$x$', '$y$'])
```

Further, if you want to graph this function *and* the function $y = 10x^2 - 4$ on the same set of axes, you can enter

```
plot(10*x^2-4, -2, 2)+plot((9*x^5-5*x^3)/(x^2+1), -2, 2, axes_labels=['$x$', '$y$'])
```

(There are two "`plot`" commands here, but you only need to specify the labeling of axes once.)

5. Graph each of the following functions separately. Put labels on the axes.
    (a) $y = (x + 1)^2$ (b) $y = 600 - x^3$ (c) $y = x^2$ (d) $y = x^2 + 1$ (e) $y = 3x^2 + x - 1$
6. Graph the function $f(x) = 1 - 2x^2$ on the interval $-1 \le x \le 1$.
7. What is the y-intercept of this graph? The graph has two x-intercepts; use algebra to find them.

You can also find an x-intercept using the computer. The idea is to **magnify** the graph near the intercept until you can determine as many decimal places in the x coordinate as you want. For a start, graph the function on the interval $0 \le x \le 1$. You should be able to see that the graph on your computer monitor crosses the x-axis somewhere around 0.7.

Regraph $f(x)$ on the interval $0.6 \leq x \leq 0.8$. You should then be able to determine that the x-intercept lies between 0.7 and 0.8. This means $x = 0.7\ldots$; that is, you know the location of the x-intercept to one decimal place of accuracy.

8. Regraph $f(x)$ on the interval $0.70 \leq x \leq 0.71$, to get two decimal places of accuracy in the location of the x-intercept. Continue this process until you have at least 4 places of accuracy. What is this x-intercept, to four places?

9. The sage `plot` command only recognizes x as a variable. If you want to plot, say, $F(w) = (w - 1)(w - 2)(w - 3)$, then you need to precede your `plot` statement with one of the form

$$\texttt{var('w')}$$

to declare w as a valid plotting variable. (Alternatively, you could just plot $F(x) = (x - 1)(x - 2)(x - 3)$; it would look the same.)

Graph each of the following functions, with axes labeled appropriately.

(a) $F(w) = (w - 1)(w - 2)(w - 3)$ (b) $Q(a) = \dfrac{1}{a^2 + 5}$ (c) $M(u) = \dfrac{u^2 - 2}{u^2 + 2}$

10. Try to execute the command

$$\texttt{plot(sqrt(x), -1, 1)}$$

What goes wrong? How can you fix it?

11. Try to execute the command

$$\texttt{plot(1/x, -1, 1)}$$

What goes wrong? How can you fix it?

12. Graph, on the same set of axes, the following three functions:

$$f(x) = 2^x, \qquad g(x) = 3^x, \qquad h(x) = 10^x.$$

Use the domain $-1 \leq x \leq 1$.

(a) Which function has the largest value when $x = -1$?
(b) Which is climbing most rapidly when $x = 0$?
(c) Magnify the picture at $x = 0$ by resetting the size of the domain to $-0.01 \leq x \leq 0.01$. Describe what you see. Estimate the slopes of the three graphs at $x = 0$.

1.7   Some Families of Functions

1.7.1   Linear Functions

Proportional changes in input and output. Suppose y is a function of x. Then there is some rule that answers the question: What is the value of y for any given x? Often, however, we start by knowing the value of y for a particular x, and the question we really want to ask is: How does y respond to *changes* in x? We are still dealing with the same function—just looking at it from a different point of view. This point of view is important; we have used it, and will continue to use it, to analyze functions (like $S(t)$, $I(t)$, and $R(t)$) that are defined by rate equations.

The way Δy depends on Δx can be simple or it can be complex, depending on the function involved. The simplest possibility is that Δy and Δx are **proportional**:

$$\Delta y = m \Delta x, \quad \text{for some constant } m. \tag{1.7}$$

(Remember: "is proportional to" means "equals a constant times.") Thus, if Δx is doubled, so is Δy; if Δx is tripled, so is Δy. A function whose input and output are related in this simple way is called a **linear function** because, as it turns out, the graph of such a function is a straight line. Let's take a moment to see why this is so.

The graph of a linear function. The graph of a linear function, call it ℓ, consists of certain points in the (x, y)-plane. Fix one point on ℓ, and call it (x_0, y_0). Now suppose (x, y) is another point on ℓ, and is Δx units away from (x_0, y_0) in the horizontal direction. Then by (1.7), (x, y) is $\Delta y = m \Delta x$ units from (x_0, y_0) in the vertical direction. Then the slope of the line connecting these two points is (Fig. 1.4)

$$\frac{\Delta y}{\Delta x} = \frac{m \Delta x}{\Delta x} = m.$$

Fig. 1.4 Points (x_0, y_0) and (x, y) on the graph of a linear function

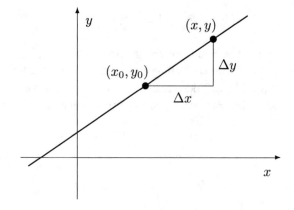

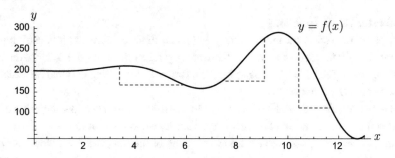

Fig. 1.5 For nonlinear functions, $\Delta y / \Delta x$ varies

Since m is a constant, this slope is the same for *all* points (x, y) on the given line ℓ. So what we have shown is that **every point on the graph of ℓ lies on the line of slope m through the point** (x_0, y_0). But this can only happen if that line *is* ℓ! We conclude that:

> **A linear function is one that satisfies $\Delta y = m \, \Delta x$;**
> **its graph is a straight line whose slope is m.**

Definition of linear function

The above definition tells us that, for a straight line, the ratio $\Delta y / \Delta x$ is *constant*. That is, this ratio does not depend on where the interval of length Δx begins or ends. Note that this is not so for functions that *do not* give straight lines. For more general functions, the ratio $\Delta y / \Delta x$ can vary depending on where the interval in question starts, and how long this interval is (Fig. 1.5).

The slope m of a linear function is sometimes called a *multiplier*, since it's what we multiply the change in x by to get the change in y.

But there's yet another interpretation of m. The equation $\Delta y = m \, \Delta x$ tells us that, if $\Delta x = 1$, then $\Delta y = m$. That is: y changes by m units for every unit change in x. For this reason, for a linear function, the slope m *is* the rate of change of y with respect to x.

All of the above is worth summarizing.

> **For a linear function satisfying $\Delta y = m \, \Delta x$, the coefficient**
> **m is the slope, multiplier, and rate of change.**

Three interpretations of m

In particular, the rate of change of a linear function is the same everywhere. Certainly this is not true of functions in general.

Formulas for linear functions

The expression $\Delta y = m \Delta x$ tells us that y is a linear function of x, but it doesn't quite give us a *formula* for y in terms of x. In fact, there are several equivalent ways to write the relation $y = f(x)$, when $f(x)$ is a linear function, in a formula, depending on what information we are given about the function.

• **The initial value form**. Here is a very common situation: we know the value of y, call it y_0, at some "initial" value of x, call it x_0. That is, we know that our line passes through the point (x_0, y_0). Suppose we also know the slope of the line, call it m. Now let (x, y) be any other point on the line. Since $\Delta y = m \Delta x$, we now that difference in x-values between these points equals m times the distance in y values, so

$$y - y_0 = m(x - x_0)$$

or, solving for y,

$$y = y_0 + m(x - x_0). \qquad (1.8)$$

See Fig. 1.4.

This is formula is what we call the initial value form the equation of a line. Note that this formula expresses y in terms a point (x_0, y_0) and a slope m. Therefore, the formula is also referred to as the **point-slope form** of the equation of a line. It may be more familiar to you with that name.

• **The interpolation form, or: two points determine a line**. If we are given that a line passes through two points (x_1, y_1) and (x_2, y_2), then we can compute that this line has slope

$$m = \frac{y_2 - y_1}{x_2 - x_1}.$$

See Fig. 1.6. But now we know the slope of the line, and we also know a point on the line— take (x_1, y_1), for example. (You could also choose (x_2, y_2); you'd get the same answer in the end. See Example 1.7.1a below.) Then we can use the point-slope form just discussed to to get the equation of the line. We have

$$y = y_1 + m(x - x_1) = y_1 + \frac{y_2 - y_1}{x_2 - x_1}(x - x_1). \qquad (1.9)$$

Notice how, once again, y is expressed in terms of the initial data—which consists of the two points (x_1, y_1) and (x_2, y_2).

The process of finding values of a quantity between two given values is called **interpolation**. Since our new expression does precisely that, it is called the interpolation formula. (Of course, it also finds values outside the given interval.) Since the initial data is a pair of points, the interpolation formula is also called the **two-point form** for the equation of a line.

• **The slope-intercept form**. This is a special case of the initial value form that occurs when the initial value x_0 of x equals 0. Then the point (x_0, y_0) lies on the y-axis, and it is frequently written in the alternate form $(0, b)$ (Fig. 1.7).

Fig. 1.6 The interpolation, or two point, form: two points determine a line

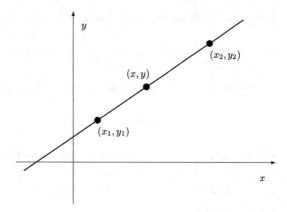

Fig. 1.7 The slope-intercept form of a line

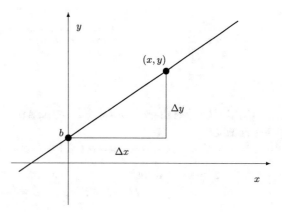

The number b is called the **y-intercept**. The equation is

$$y = mx + b. \tag{1.10}$$

This equation is called the slope-intercept form of the equation for our line. In some sense it's the simplest-looking form, though it's not always the most useful.

Example 1.7.1 Find an equation for each line with the given properties. Then put your equation into slope-intercept form.

(a) The line through $(-1, 2)$ and $(-6, 4)$.
(b) The line through $(3, 1)$, and such that every two units change in x produces -4 units change in y.
(c) The line through $(0, 2)$ with slope $2/3$.

Solution. (a) We use the interpolation, or two-point, formula (1.9); we find that

$$y = y_1 + \frac{y_2 - y_1}{x_2 - x_1}(x - x_1) = 2 + \frac{4 - 2}{-6 - (-1)}(x - (-1)) = 2 + \frac{2}{-5}(x - (-1)) = 2 - \frac{2}{5}(x + 1).$$

To put this into slope-intercept form, we just simplify:

$$y = 2 - \frac{2}{5}(x + 1) = 2 - \frac{2}{5}x - \frac{2}{5} = -\frac{2}{5}x + \frac{8}{5}.$$

(So the slope is $m = -2/5$; the y-intercept is $b = 8/5$.) Note that, for this computation, we made a choice of which point to think of as (x_1, y_1), and which to think of as (x_2, y_2). But, had we made the other choice, we would (in general; not just in this example) have obtained the same answer:

$$y = y_1 + \frac{y_2 - y_1}{x_2 - x_1}(x - x_1) = 4 + \frac{2 - 4}{-1 - (-6)}(x - (-6)) = 2 + \frac{-2}{5}(x - (-6)) = 4 - \frac{2}{5}(x + 6)$$

$$= 4 - \frac{2}{5}x - \frac{12}{5}x = -\frac{2}{5}x + \frac{8}{5}.$$

(b) This line has slope $m = -4/2 = -2$, and therefore, by the point-slope formula (1.8), has equation

$$y = y_0 + m(x - x_0) = 1 + (-2)(x - 3) = 1 - 2(x - 3)$$

or, in slope-intercept form,

$$y = 1 - 2x + 6 = -2x + 7.$$

(c) The equation is

$$y = \frac{2}{3}x + 2.$$

The upshot of the above discussions is that, when the change in one quantity is proportional to the change in another, then the relation between these quantities follows a *linear model*. Here's an example that gives a context for these ideas.

Example 1.7.2 Measurements show that the length L of a metal bar increases in proportion to the increase in temperature T. That is, $\Delta L = m\Delta T$ for some number m.

An aluminum bar that is exactly 100 in. long when the temperature is $40\,°F$ becomes 100.0052 in. long when the temperature increases to $80\,°F$.

(a) How long is the bar when the temperature is $60\,°F$? $100\,°F$?
(b) What is the rate of change m that connects an increase in length ΔL to an increase in temperature ΔT?
(c) How long will the bar be when $T = 0\,°F$?

(d) Express L as a linear function of T.

(e) What temperature change would make $L = 100.01$ in.?

(f) What is the rate of change μ that connects an increase in temperature ΔT to an increase in length ΔL? (The symbol μ is the Greek letter *mu*.)

(g) Express T as a linear function of L.

Solution. (a) We are told that the change in length is proportional to the change in temperature. A $40°$ F increase in temperature—from $40°$F to $80°$F—produces a length change of

$$100.0052 - 100 = 0.0052$$

in (inches). So half that temperature increase—from $40°$F to $60°$F—will produce half the increase in length, meaning an increase of $0.0052/2 = 0.0026$ in. So at $60°$F, the length of the bar will be $100 + 0.0026 = 100.0026$ in.

(b) In part (a) we observed that, when $\Delta T = 40°$F, we have $\Delta L = 0.0052$ in. So the equation $\Delta L = m\,\Delta T$ satisfies 0.0052 in $= m \cdot 40°$F. Solving for m gives

$$m = \frac{0.0052 \text{ in}}{40° \text{ F}} = 0.00013 \,\frac{\text{in}}{°\text{F}}.$$

That is, $\Delta L = 0.00013\,\Delta T$.

(c) If $\Delta T = -40°$F, then $\Delta L = 0.00013 \cdot (-40) = -0.0052$ in, which implies a length of $L = 100 + (-0.0052) = 99.9948$ in, when $T = 0°$F. (The bar shrinks just as much, for a temperature *drop* of $40°$F, as it expands for a temperature *gain* of $40°$F.)

(d) The line expressing L in terms of T has slope $m = 0.00013$ and L-intercept $b = 99.9948$, and therefore, by the slope-intercept formula (1.10), has equation

$$L = 0.00013T + 99.9948. \tag{1.11}$$

(e) We solve $100.01 = 0.00013T + 99.9948$ for T, to get

$$T = \frac{100.01 - 99.9948}{0.00013} = 116.9231° \text{ F}.$$

(f) Solving $\Delta L = 0.00013\,\Delta T$ for ΔT gives $\Delta T = \Delta L/0.00013 = 7692.3077\,\Delta L$. So $\mu = 7692.3077$ degrees Fahrenheit per inch.

(g) We can use the point-slope formula (1.8). Or we can solve (1.11) for T, to get

$$T = \frac{L - 99.9948}{0.00013} = 7692.3077L - 769190.7692.$$

The equation

$$\Delta y = m\Delta x$$

of a linear function, with m being the rate of change of the function, should look familiar. We've seen a situation where

change in one variable = rate of change, times change in another variable,

or more precisely,

change in one variable ≈ rate of change, times change in another variable,

before. Namely: we had the "prediction equation" $\Delta Q \approx Q' \Delta t$ (see Eq. (1.2)), which we used as a central part of Euler's method.

For linear functions, the rate of change is *constant*; it is for this reason that we can use "=" instead of "≈" in the prediction equation, in cases where output depends in a linear fashion on input. Certainly this is *not* the case in our above SIR model: neither S nor I nor R is a linear function of a time. Euler's method amounts to approximating these functions with functions that are linear "in pieces." More specifically, Euler's method approximates a function with one that is linear on intervals of length Δt, and whose slope on each of these intervals is computed using rate equation information.

1.7.2   The Circular Functions

Graphing packages "know" the familiar functions of trigonometry. Trigonometric functions are qualitatively different from the functions in the preceding problems. Those functions are defined by algebraic formulas (that is, formulas involving only addition, subtraction, multiplication, division, exponentiation, and roots), so they are called **algebraic functions**. The trigonometric functions are defined by explicit "recipes," but *not* by algebraic formulas; they are called **transcendental functions**. For calculus, we usually use the definition of the trigonometric functions as **circular functions**. This definition begins with the unit circle, meaning the circle of radius 1 centered at the origin. Given the input number t, trace an arc of length $|t|$ along the circle, starting from the point $(1, 0)$. If t is positive, trace the arc counterclockwise; if t is negative, trace it clockwise. Suppose this motion takes you to a point P on the circle. The circular (or trigonometric) functions $\cos(t)$ and $\sin(t)$ are then defined as the coordinates of the point P (Fig. 1.8),

$$P = (\cos(t), \sin(t)).$$

The other trigonometric functions are defined in terms of the sine and cosine:

$$\tan(t) = \sin(t)/\cos(t), \qquad\qquad \sec(t) = 1/\cos(t),$$
$$\cot(t) = \cos(t)/\sin(t), \qquad\qquad \csc(t) = 1/\sin(t).$$

Notice that when t is a positive acute angle, the circle definition agrees with the right triangle definitions of the sine and cosine (see Fig. 1.9).

Fig. 1.8 Definitions of $\cos(t)$ and $\sin(t)$ as coordinates of points on the unit circle

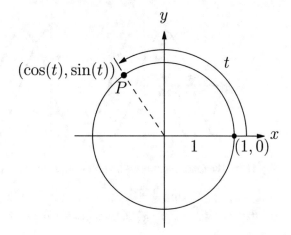

Fig. 1.9 $\cos(t)$ and $\sin(t)$ as coordinates of points on the unit circle, and as sidelengths of a right triangle

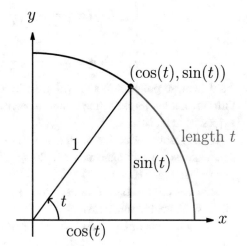

$$\sin(t) = \frac{\text{opposite}}{\text{hypotenuse}} \quad \text{and} \quad \cos(t) = \frac{\text{adjacent}}{\text{hypotenuse}}.$$

However, the circle definitions of the sine and cosine have the important advantage that they produce functions whose domains are the set of *all* real numbers. (What are the domains of the tangent, secant, cotangent and cosecant functions?)

In moving to the point P, with coordinates $(\cos(t), \sin(t))$, as just described, we are sweeping out an angle centered at the origin. We say that this angle has **radian measure** t. In calculus, angles are always measured in radians. To convert between radians and degrees, notice that the circumference of the unit circle is 2π, so the radian measure of a semi-circular arc is half of this, and thus we have

$$\pi \text{ radians } = 180 \text{ degrees.}$$

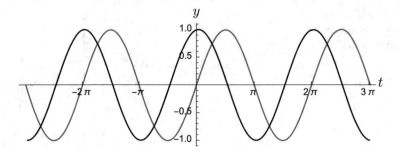

Fig. 1.10   The functions $y = \cos(t)$ (in black) and $y = \sin(t)$ (in red)

In other words, 1 radian $= 180/\pi$ degrees and 1 degree $= \pi/180$ radians. So, for example,

$$60 \text{ degrees} = 60 \cdot \frac{\pi}{180} \text{ radians} = \frac{\pi}{3} \text{ radians},$$

and

$$\frac{-7\pi}{4} \text{ radians} = \frac{-7\pi}{4} \cdot \frac{180}{\pi} \text{ degrees} = -315 \text{ degrees}.$$

As the course progresses, you will see why radians are used rather than degrees—it will turn out that the formulas important to calculus take their simplest form when angles are expressed in radians.

If one graphs the cosine of an angle on the vertical axis, against the radian measure of that angle on the horizontal axis, and does the same for the sine of an angle, then one gets a picture like Fig. 1.10.

Remark on notation. It is common to simply write $\cos t$ for $\cos(t)$, and similarly for other trigonometric functions.

Moreover, one often writes $\cos^2(t)$, or (less frequently) $\cos^2 t$, for $(\cos(t))^2$, and similarly for any powers of any of the circular functions. Some care is required; for example, $\sin^{23} 24t \cos^{24} 23t$ might be a bit hard to read. And things like $\sin 2x - 7$ are flat out ambiguous. We'll use these abbreviated notations only in simple circumstances, when the context makes the symbols clear.

1.7.3   Functions Proportional to Their Rates of Change

Earlier in this section, we discussed situations where a change Δy in one quantity is proportional to the change Δx in another. A somewhat analogous, though mathematically quite different, situation is where the *rate of change* of a quantity is proportional to that quantity itself.

Consider a human population as an example. If a city of 100,000 persons is increasing at the rate of 1,500 persons per year, we might expect a similar city of 200,000 persons to be increasing at the rate of 3,000 persons per year. That is, we might expect the rate P' of growth of a population to be proportional to the size P of the population. In symbols,

$$P' = kP$$

A quantity proportional to its rate of change

In the present circumstance, we would have $k > 0$, since the population is growing rather than shrinking.

Note that solving the above equation for k gives

$$k = \frac{P'}{P}.$$

So k is the growth rate divided by the number of persons; that is, k is the growth rate *per person*. For this reason, k is often called the *per capita* growth rate. ("Per capita" literally means "per head.")

In particular, the units of k are those of P' divided by those of P. So, for example, if P is measured in persons and time t in years, then the units for k are

$$\frac{\text{persons/year}}{\text{person}}, \quad \text{or simply 1/year (or year}^{-1}\text{)}.$$

More generally, if P is any quantity proportional to its rate of change, then the constant $k = P'/P$ can be considered a *per unit* growth rate. And the units of k are the inverse to the units of the independent variable in question.

In a later chapter, we will see that the solution $P(t)$ to a rate equation of the form $P' = kP$ is an *exponential* function of t. Exponential functions, as we'll see, behave very differently from linear functions.

For now, we study such rate equations using Euler's method.

Example 1.7.3 In 1985, the per capita growth rate in Poland was 9 persons per year per thousand persons. Assuming that the population of Poland grows in the manner described above:

(a) Let P denote the population of Poland. Write a rate equation for P' in terms of P.

(b) In 1985, the population of Poland was estimated to be 37.5 million persons. What was the net growth rate P' (as distinct from the *per capita* growth rate) in 1985?

(c) Use Euler's method, with $\Delta t = 3$ months, to estimate the population of Poland in 1986.

(d) About how long did it take the population to increase by one person in Poland in 1985?

Solution. (a) We have $P' = kP$. But the given information tells us that, in 1985, $P'/P = 9/1000 = 0.009$, so $k = 0.009$, so

$$P' = 0.009P.$$

Here, P is in persons, P' in persons per year, and k in persons per year per year, or year^{-1}.

(b) In 1985,

$$P' = 0.009 \cdot 37.5 = 0.3375$$

million persons per year.

(c) In years, we have $\Delta t = 1/4 = 0.25$. We take 1985 as $t = 0$; then Euler's method gives

$$
\begin{aligned}
P(1/4) &= P(0) + \Delta P \approx P(0) + P'(0)\Delta t \\
&= P(0) + (kP(0))\Delta t = 37.5 + (0.009 \cdot 37.5 \cdot 0.25) = 37.5844, \\
P(1/2) &= P(1/4) + \Delta P \approx P(1/4) + P'(1/4)\Delta t \\
&\approx P(1/4) + (kP(1/4))\Delta t = 37.5844 + (0.009 \cdot 37.5844 \cdot 0.25) = 37.6690, \\
P(3/4) &= P(1/2) + \Delta P \approx P(1/2) + P'(1/2)\Delta t \\
&\approx P(1/2) + (kP(1/2))\Delta t = 37.6690 + (0.009 \cdot 37.6690 \cdot 0.25) = 37.7538, \\
P(1) &= P(3/4) + \Delta P \approx P(3/4) + P'(3/4)\Delta t \\
&\approx P(3/4) + (kP(3/4))\Delta t = 37.7538 + (0.009 \cdot 37.7538 \cdot 0.25) = 37.8387
\end{aligned}
$$

million persons.

(d) We have $\Delta P \approx P'\Delta t$. So by part (b) we find that, in 1985, a ΔP of one person corresponds to an elapsed time of

$$\Delta t \approx \frac{\Delta P}{P'} = \frac{1}{0.3375 \cdot 10^6} = 2.9630 \cdot 10^{-6}$$

years, or

$$2.9630 \cdot 10^{-6} \ \text{yr} \cdot 365 \ \frac{\text{day}}{\text{yr}} \cdot 24 \ \frac{\text{h}}{\text{day}} \cdot 60 \ \frac{\text{min}}{\text{h}} \cdot 60 \ \frac{\text{s}}{\text{min}} = 93.4412 \ \text{s}.$$

1.7.4 Exercises

1.7.4.1 Part 1: Linear Functions and Graphs

1. Find an equation for each line with the given properties. Then put your equation into slope-intercept form.

   (a) The line through $(0, -2)$ with slope 5.
   (b) The line through $(3, 7)$ and $(6, -2)$.

(c) The line through $(3, 1)$, and such that every decrease of one unit in y produces a three unit increase in x.

2. You should be able to answer all parts of this problem without ever finding the equations of the functions involved.

(a) Suppose $y = f(x)$ is a linear function with multiplier $m = 3$. If $f(2) = -5$, what is $f(2.1)$? $f(2.0013)$? $f(1.87)$? $f(922)$?

(b) Suppose $y = G(x)$ is a linear function with multiplier $m = -2$. If $G(-1) = 6$, for what value of x is $G(x) = 8$? $G(x) = 0$? $G(x) = 5$? $G(x) = 491$?

(c) Suppose $y = h(x)$ is a linear function with $h(2) = 7$ and $h(6) = 9$. What is $h(2.046)$? $h(2 + a)$?

3. Sketch, using a computer, the graph of each of the following linear functions. Label each axis. For each line that you draw, indicate (i) its slope; (ii) its y-intercept; (iii) its x-intercept (where it crosses the x-axis).

(a) $y = -\frac{1}{2}x + 3$ (b) $5x + 3y = 12$ (c) $y = (2x - 7)/3$

4. Graph the function $f(x) = 0.6x + 2$ on the interval $-4 \le x \le 4$.

(a) What is the y-intercept of this graph? What is the x-intercept?

(b) Read from the graph the value of $f(x)$ when $x = -1$ and when $x = 2$. What is the difference between these output values? What is the difference between the x values? According to these differences, what is the slope of the graph? According to the *formula*, what is the slope?

1.7.4.2 Part 2: Linear Models

5. In Colorado there is a sales tax of 2.9%. The tax T, in dollars, is proportional to the price P of an object, also in dollars. The constant of proportionality is $k = 2.9\% = 0.029$. Write a formula that expresses the sales tax as a linear function of the price, and use your formula to compute the tax on an LED television that costs $689.00 and a pair of earbuds that costs $29.99.

6. Suppose $W = 213 - 17Z$. How does W change when Z changes from 3 to 7; from 3 to 3.4; from 3 to 3.02? Let ΔZ denote a change in Z and ΔW the change thereby produced in W. Is $\Delta W = m\,\Delta Z$ for some constant m? If so, what is m?

7. a. In the following table, q is a linear function of p. Fill in the blanks in the table.

p	-3	0		7	13		π
q	7		4	1		0	

b. Find a formula to express Δq as a function of Δp, and another to express q as a function of p.

8. **Thermometers.** There are two scales in common use to measure the temperature, called **Fahrenheit degrees** and the **Celsius degrees**. Let F and C, respectively, be the temperature on each of these scales. Each of these quantities is a linear function of the other; the relation between them in determined by the following table:

physical measurement	C	F
freezing point of water	0	32
boiling point of water	100	212

(a) Which represents a larger change in temperature, a Celsius degree or a Fahrenheit degree?

(b) How many Fahrenheit degrees does it take to make the temperature go up one Celsius degree? How many Celsius degrees does it take to make it go up one Fahrenheit degree?

(c) What is the multiplier m in the equation $\Delta F = m\Delta C$? What is the multiplier μ in the equation $\Delta C = \mu\Delta F$? What is the relation between μ and m?

(d) Express F as a linear function of C. (We have already done this in an earlier section. Try to do it again "from scratch," using only the previous parts of this exercise.)

(e) Express C as a linear function of F.

(f) Is there any temperature that has the same reading on the two temperature scales? What is it? Does the temperature of the air ever reach this value? Where?

9. **The Greenhouse Effect.** The concentration of carbon dioxide (CO_2) in the atmosphere is increasing. The concentration is measured in parts per million (ppm). Records kept at the Mauna Loa Observatory in Hawaii show an increase of 0.8 ppm per year during the 1960s.

(a) At that rate, how many years does it take for the concentration to increase by 5 ppm; by 15 ppm?

(b) At the beginning of 1960, the concentration was about 3167 ppm. Assuming that the linear relationship continues, what would it be at the beginning of 1980; at the beginning of 2000; at the beginning of 2020?

(c) Draw a graph that shows CO_2 concentration as a function of time since 1960, assuming the linear relationship described above. (Do this by hand or on a computer.) Label everything clearly.

(d) The *actual* CO_2 concentration at the Mauna Loa Observatory was 339 ppm at the beginning of 1980, 370 ppm at the beginning of 2000, and 414 ppm at the beginning of 2020. Plot these values on your graph, and compare them to your calculated values.

(f) Using the actual concentrations in 2000 and 2020, calculate a new rate of increase in concentration. Using that rate, estimate what the CO_2 concentration will be in 2030. Do you think your projection is realistic? Why or why not?

(g) Using the rate of 0.8 ppm per year that held during the 1960s, determine how many years before 1960 there would have been *no* carbon dioxide at all in the atmosphere.

10. **Falling Bodies**. In the simplest model of the motion of a falling body, the velocity increases in proportion to the duration of time that the body has been falling. If the velocity is given in feet per second, measurements show the constant of proportionality is approximately 32.

(a) A ball is falling at a velocity of 40 ft/s after 1 s. How fast is it falling after 3 s?

(b) Express the change in the ball's velocity Δv as a linear function of the change in time Δt.

(c) Express v as a linear function of t. Assume that the ball was simply dropped, so that its initial velocity was zero.

The model can be expanded to keep track of the *distance* that the body has fallen. If the distance d is measured in feet, the units of d' are feet per second; in fact, $d' = v$. So the model describing the motion of the body is given by the rate equations

$$d' = v \quad \text{feet per second;}$$
$$v' = 32 \quad \text{feet per second per second.}$$

(d) At what rate is the distance increasing after 1 s? After 2 s? After 3 s?

(e) Is d a linear function of t? Explain your answer.

1.7.4.3 Part 3: Circular, or Trigonometric, Functions

11. Convert the following radian measures to degrees:

(a) -7π (b) $\dfrac{2\pi}{3}$ (c) $-\dfrac{5\pi}{4}$ (d) $\dfrac{5\pi}{12}$ (e) $\dfrac{\pi}{6}$ (f) 11

12. Convert the following degree measures to radians:

(a) 45   (b) -180 (c) 36   (d) -30 (e) 270   (f) -11

13. Consider the function $y = \tan(x) = \sin(x)/\cos(x)$.

(a) What is the largest interval containing the origin on which this function is defined? Please explain.

(b) Use a computer to graph $y = \tan(x)$ on this interval. If you encounter some problems, think about what's happening at the endpoints of your interval, and then try graphing over a slightly smaller interval.

14. There is a well-known "trigonometric limit formula" that says the following: the smaller (that is, closer to zero) the number t gets, the closer the ratio

$$\frac{\sin(t)}{t}$$

gets to 1. In symbols, we write

$$\lim_{t \to 0} \frac{\sin(t)}{t} = 0.$$

(See Chap. 2 for more on the notion of limits.)

Explain what this means in terms of various lengths depicted in Fig. 1.9. Do you think that Fig. 1.9 makes this limit formula plausible?

15. Another well-known trigonometric limit formula that says the following: the smaller (that is, closer to zero) the number t gets, the closer the ratio

$$\frac{\cos(t) - 1}{t}$$

gets to 0. In symbols, we write

$$\lim_{t \to 0} \frac{\cos(t) - 1}{t} = 0.$$

Explain what this limit formula means in terms of various lengths depicted in Fig. 1.9. Do you think that Fig. 1.9 makes this limit formula plausible?

16. Explain geometrically, in terms of Fig. 1.8 or Fig. 1.9, why the functions $\cos(t)$ and $\sin(t)$ are 2π-*periodic*, meaning

$$\cos(t + 2\pi) = \cos(t) \quad \text{and} \quad \sin(t + 2\pi) = \sin(t)$$

for any real number t. Hint: think about sweeping out an arc of length t, on the unit circle, versus sweeping out an arc of length $t + 2\pi$.

17. A well-known "trigonometric identity" states that

$$\cos^2(t) + \sin^2(t) = 1$$

for any real number t. Explain geometrically why this identity is true. Hint: Fig. 1.8 or Fig. 1.9.

18. Show that

$$\sec^2(t) = 1 + \tan^2(t)$$

for any real number t. Hint: start with the result of the previous exercise, and divide both sides by the appropriate quantity.

19. The following exercise lets you review the trigonometric functions and explore them using computer graphing. (Note: in Sage, $\cos(x)$ and $\sin(x)$ are simply coded as `cos(x)` and `sin(x)`.)

(a) Graph the function $f(x) = \sin(x)$ on the interval $-2 \leq x \leq 10$.

(b) What are the x-intercepts of $\sin(x)$ on the interval $-2 \leq x \leq 10$? Determine them to two decimal places accuracy.

(c) What is the largest value of $f(x)$ on the interval $-2 \leq x \leq 10$? Which value of x makes $f(x)$ largest? Determine x to two decimal places accuracy.

(d) Regraph $f(x)$ on the small interval $-0.01 \leq x \leq 0.01$. Describe what you see. Estimate the slope of this graph at $x = 0$.

(e) Graph the function $f(x) = \cos(x)$ on the domain $0 \leq x \leq 14$. On the same set of axes, graph the function $g(x) = \cos(2x)$.

(f) How far apart are the x-intercepts of $f(x)$? How far apart are the x-intercepts of $g(x)$?

(g) The graph of $g(x)$ has a pattern that repeats. How wide is this pattern? The graph of $f(x)$ also has a repeating pattern; how wide is *it*?

(f) Compare the graphs of $f(x)$ and $g(x)$ to one another. In particular, can you say that one of them is a stretched or compressed version of the other? Is the compression (or stretching) in the vertical or the horizontal direction?

(g) Construct a *new* function $h(x)$ whose graph is the same shape as the graph of $g(x) = \cos(2x)$, but make the graph of $h(x)$ twice as tall as the graph of $g(x)$. [A suggestion: either deduce what $h(x)$ should be, or make a guess. Then test your choice on the computer. If your choice doesn't work, think how you might modify it, and then test your modifications the same way.] Graph h on the same set of axes as you did f and g.

20. The aim here is to find a solution to the equation $\sin x = \cos(3x)$. There is no purely *algebraic* procedure to solve this equation. Because the sine and cosine are not defined by *algebraic* formulas, this should not be particularly surprising. (Even for algebraic equations, there are only a few very special cases for which there are formulas like the quadratic formula.)

(a) Graph the two functions $f(x) = \sin(x)$ and $g(x) = \cos(3x)$ together on the interval $0 \leq x \leq 1$.

(b) Find a solution of the equation $\sin(x) = \cos(3x)$ that is accurate to six decimal places.

(c) Find *another* solution of the equation $\sin(x) = \cos(3x)$, accurate to four decimal places. Explain how you found it.

1.7.4.4 Part 3: Functions Proportional to Their Rates of Change

For the exercises in this part, please refer to the subsection "Functions proportional to their rates of change," and to Example 1.7.3, above.

21. **Afghanistan**. In 1985 the per capita growth rate in Afghanistan was 21.6 persons per
    year per thousand.

    (a) Let A denote the population of Afghanistan. Write the equation that governs the
        growth rate A' of A.
    (b) In 1985 the population of Afghanistan was estimated to be 15 million persons. What
        was the net growth rate A' in 1985?
    (c) Comparing part (b) of this exercise with part (b) of Example 1.7.3, comment on
        the following assertion: When comparing two countries, the one with the larger per
        capita growth rate will have the larger net growth rate.
    (d) Use Euler's method, with $\Delta t = 4$ months, to estimate the population of Afghanistan
        in 1986.
    (e) About how long did it take the population to increase by one person in Afghanistan
        in 1985? How does this compare with part (d) of Example 1.7.3?

22. **Bacterial Growth**. A colony of bacteria on a culture medium grows at a rate propor-
    tional to the present size of the colony. When the colony weighed 32 grams it was
    growing at the rate of 0.79 grams per hour.

    (a) Write an equation that links the growth rate B' to the size B of the population. Hint:
        the per capita growth rate k is *not* equal to 0.79. Rather, k can be found by plugging
        the given information into your rate equation.
    (b) What are the units for your per capita growth rate in part (a) above?
    (c) Use Euler's method with $\Delta t = 1/2$ h to estimate B after one hour.

23. **Radioactivity**. In radioactive decay, radium slowly changes into lead. If one sample of
    radium is twice the size of a second lump, then the larger sample will produce twice
    as much lead as the second in any given time. In other words, the rate of decay is
    proportional to the amount of radium present. *Decay* means *decrease* in the amount
    present, so if we denote this amount by R, then we have

    $$R' = -kR$$

    (note the minus sign), where k is a positive parameter, called the (*per unit*) *decay rate*.

    (a) Measurements show that 1 g of radium decays into lead at the rate of 1/2337 grams
        per year. (That is, $R' = -1/2337$ grams per year when $R = 1$ gram.) Using this
        information, evaluate k. That is, supply a numerical value for k, and also state what
        the appropriate units for k are.
    (b) Using a step size of 10 years, use Euler's method to estimate how much radium
        remains in a 0.072 g sample after 40 years.

24. **Newton's Law of Cooling**. Suppose a cup of hot coffee is brought into a room with constant temperature $20\,°C$. The coffee will cool off, and it will cool off *faster* when the temperature difference between the coffee and the room is greater. The simplest assumption we can make is that the rate of cooling is proportional to this temperature difference. This is called Newton's Law of Cooling.

Let C denote the temperature of the coffee, in $°C$, and C' the rate at which it is cooling, in $°C$ per minute. The new element here is that C' is proportional, not to C, but to the *difference* between C and the room temperature of $20\,°C$.

(a) Write an equation that relates C' and C. It will contain a proportionality constant k. How did you indicate that the coffee is *cooling* and not *heating up*?

(b) When the coffee is at $90\,°C$ it is cooling at the rate of $7\,°C$ per minute. What is k ?

(c) At what rate is the coffee cooling when its temperature is $50°C$?

(d) Estimate how long it takes the temperature to fall from $90\,°C$ to $50\,°C$. Then make a better estimate, and explain why it is better.

25. **Moore's Law.** According to *Moore's Law* (formulated by Gordon Moore in 1965), the maximum number N of transistors that can be fit on a microchip increases at a rate proportional to N.

(a) Write an equation that relates N' and N. It will contain a proportionality constant k. What about your equation tells you that N is *increasing* and not *decreasing*?

(b) In the year 2018, N was equal to 23,600,000,000 (23.6 billion, or $2.36 \cdot 10^{10}$), and was increasing at the rate of 80 billion transistors per year. What is k?

(c) Using Euler's method with $\Delta t = 1$ year, estimate the number of chips that could fit on a microchip in 2021. If you are interested, do some research to compare this estimate with the actual number.

1.8 Summary

Natural phenomena like epidemics can frequently be described by **mathematical models**. In particular, they can often be described by **dynamical systems**, where the **rates of change** of the quantities in question are described mathematically, through **rate equations**.

Information about the rate of change of a given quantity (that is, information about "how fast we're going") allows us to predict how this quantity will evolve (that is, it allows us to predict "how far we'll get in a given amount of time"). This is the **prediction principle**. This prediction will, in general, only be approximate, since the rate of change is typically itself changing. But over short enough intervals of time, over which the rate of change does not change much, the prediction should be a relatively good one.

We can then apply the prediction principle iteratively, over many short intervals in succession, to ultimately predict farther out into the future. This process amounts to **Euler's method**. Because many iterations may be required to get a "good" approximation, Euler's method is computationally intensive. But fortunately, using **computer programs** with built-in **loops**, we can perform the Euler's method computations rapidly and efficiently.

The quantities S, I, and R that we studied, in the context of epidemics, are **functions** of time, in that each input time value t will, in theory, uniquely determine the corresponding output value $S(t)$, $I(t)$, or $R(t)$. We do not have **formulas** for S, I, or R as functions of t, as we do for many other functions. Still, we can understand S, I, and R by drawing (approximate) **graphs** of them, or by recording **tables** of their (approximate) values. Here again, **computer code** can be extremely helpful.

Linear functions are particularly easy to understand because they have **constant rates of change:** if $y = f(x)$ is a linear function, then $\Delta y = m \Delta x$ for some constant m. When we implement Euler's method, we are approximating nonlinear functions by functions built up of consecutive linear "pieces," each piece being defined over an interval of length Δt, where Δt is our stepsize.

In addition to functions with constant rates of change, we often encounter functions with **constant per unit rates of change**. By such a function P, we mean one such that P'/P is constant. We'll see in Chap. 3 that such functions are **exponential** in nature.

For both linear and exponential functions, then, certain kinds of rates of change are constant. Further, linear and exponential functions are both ubiquitous in nature and in math. But the similarities more or less end there.

The Derivative

2

In studying SIR and other phenomena in Chap. 1, we contented ourselves with an intuitive, or heuristic, understanding of what a rate of change actually *is*. That is, we generally avoided defining **rate of change** in a mathematically precise way.

In this chapter, we will provide such a definition. Two such definitions, actually—one of an *average* rate of change, also known as a **difference quotient**, and one of the *instantaneous* rate of change, also known as the **derivative**.

We've encountered both notions in the previous chapter. We now investigate these ideas more formally, and in greater depth.

2.1 Rates of Change

By an *average rate of change* of an output y with respect to an input x, we mean the net change Δy in output divided by the corresponding change Δx in input. That is, we mean the quantity $\Delta y/\Delta x$. In preceding discussions, we have generally taken our input variable to be time t, but other independent variables are possible, as illustrated in the following example.

Example 2.1.1 (*Water density*) Under appropriate atmospheric conditions, the density of water, as a function of water temperature, may be modeled fairly well by the formula

$$D(C) = 999.973 - 0.008(C - 4.06)^2,$$

where C is temperature in degrees Celsius (°C) and D is density in kilograms per cubic meter (kg/m^3). This formula holds reasonably well for C between about 0 and 8 °C.

© The Author(s), under exclusive license to Springer Nature Switzerland AG 2023
E. Stade and E. Stade, *Calculus: A Modeling and Computational Thinking Approach*,
Synthesis Lectures on Mathematics & Statistics,
https://doi.org/10.1007/978-3-031-24681-4_2

(a) Find the average rate of change of D with respect to C, over each of the following
   intervals (of temperature values, in °C): [1, 2], [1, 1.1], [1, 1.01], [1, 1.001], [1, 1.0001],
   and [1, 1.00001]. What are the appropriate units for these rates of change?
(b) Repeat part (a), but this time with these temperature intervals: [2, 3], [2, 2.1], [2, 2.01],
   [2, 2.001], [2, 2.0001], and [2.2.00001].
(c) Can we make sense of "the *instantaneous* rate of change of water density with respect
   to temperature, *at* 1 °C"? If so, what numerical value might we give this instantaneous
   rate of change? Answer the same questions for $C = 2$ °C.

Solution. (a) Over the interval [1, 2], the average rate of change is

$$\frac{\Delta D}{\Delta C} = \frac{D(2) - D(1)}{2 - 1} \frac{\text{kg/m}^3}{\text{°C}} = \frac{999.973 - 0.008(2 - 4.06)^2 - (999.973 - 0.008(1 - 4.06)^2)}{1} \frac{\text{kg/m}^3}{\text{°C}}$$

$$= \frac{999.93905 - 999.89809}{1} \frac{\text{kg/m}^3}{\text{°C}} = 0.04096 \frac{\text{kg/m}^3}{\text{°C}}.$$

Over [1, 1.1], the average rate of change is

$$\frac{\Delta D}{\Delta C} = \frac{D(1.1) - D(1)}{1.1 - 1} \frac{\text{kg/m}^3}{\text{°C}} = \frac{999.902907 - 999.898091}{0.1} \frac{\text{kg/m}^3}{\text{°C}} = 0.04186 \frac{\text{kg/m}^3}{\text{°C}}.$$

In a similar manner, we find the remaining entries of the following table. (All entries in the
first row are in °C; all entries in the second row are in $(\text{kg/m}^3)/\text{°C}$.)

Interval	[1, 2]	[1, 1.1]	[1, 1.01]	[1, 1.001]	[1, 1.0001]	[1, 1.00001]
$\Delta D/\Delta C$	0.04096	0.04186	0.04888	0.04895	0.04896	0.04896

(b) We need to compute

$$\frac{\Delta D}{\Delta C} = \frac{D(C) - D(2)}{C - 2}$$

for various values of C, getting closer and closer to 2 (namely, $C = 3$, 2.1, 2.01, 2.001,
2.0001, 2.00001). The computations are much as in part (a), and are summarized in the
table below.

Interval	[2, 3]	[2, 2.1]	[2, 2.01]	[2, 2.001]	[2, 2.0001]	[2, 2.00001]
$\Delta D/\Delta C$	0.02496	0.03216	0.03288	0.03295	0.03296	0.03296

(c) In part (a), we computed the average rate of change $\Delta D/\Delta C$ over shorter and shorter
temperature intervals $[1, 1 + \Delta C]$. We might think of the *instantaneous* rate of change of
D with respect to C, at $C = 1$, as "what happens to these average rates of change as the
intervals $[1, 1 + \Delta C]$ become *infinitesimally* short." Now observe from our computations
in part (a) that, the shorter our interval $[1, 1 + \Delta C]$—that is, the smaller our ΔC—the more

$\Delta D/\Delta C$ appears to zero in on 0.04896. So we might say that the instantaneous rate of change of D with respect to C, at $C = 1$, is about 0.04896 $(\text{kg/m}^3)/^\circ\text{C}$.

Similarly, according to our computations in part (b), we might say that the instantaneous rate of change of D with respect to C, at $C = 2$, is about 0.03296 $(\text{kg/m}^3)/^\circ\text{C}$.

As the above example indicates, average rates of change may be expressed mathematically in terms of functions.

Definition 2.1.1 Consider a function $y = f(x)$. Suppose x changes from a point $x = a$ to a point $x = a + \Delta x$ (so that x changes by Δx). Then the corresponding change in y is

$$\Delta y = f(a + \Delta x) - f(a),$$

and we define the **average rate of change of f, or of y, from $x = a$ to $x = a + \Delta x$**, to be the **difference quotient**

$$\frac{\Delta y}{\Delta x} = \frac{f(a + \Delta x) - f(a)}{\Delta x}. \tag{2.1}$$

Part (c) of Example 2.1.1 points to a crucial idea: the interpretation of an instantaneous rate of change as a *limit* of average rates of change, as we average over shorter and shorter intervals. That is: suppose we can somehow ascribe an actual mathematical value to "what happens to the average rate of change (2.1) as Δx shrinks to zero." Then we should call this value "the instantaneous rate of change of $f(x)$ at $x = a$."

Another name for such an instantaneous rate of change is **derivative.** The formal definition is as follows.

Definition 2.1.2 Given a function $y = f(x)$ and a point $x = a$, we define the **instantaneous rate of change**, or **derivative, of $y = f(x)$ at $x = a$**, denoted $f'(a)$, to be "what happens to the average rate of change (2.1) as Δx shrinks to zero." In symbols,

$$f'(a) = \lim_{\Delta x \to 0} \frac{\Delta y}{\Delta x} = \lim_{\Delta x \to 0} \frac{f(a + \Delta x) - f(a)}{\Delta x}, \tag{2.2}$$

where the notation "$\lim_{\Delta x \to 0}$" is pronounced "the limit, as Δx approaches zero." This definition applies whenever the limit in question exists.

The above definition is only "formal" insofar as the notion of "limit" is formal. We will content ourselves with a *working* notion of "limit"—a notion that will allow us to *compute* some derivatives, but will also provide some insight into how and when a derivative might *fail* to exist. We'll return to the latter issue in the next section. In the meantime, here are some computations that work.

Example 2.1.2 Let $f(x) = x^2$. Find:

(a) The average rate of change of $f(x)$ with respect to x, from $x = 3$ to $x = 3.1$, and from $x = 3$ to $x = 3.01$;
(b) The average rate of change of $f(x)$ from $x = 3$ to $x = 3 + \Delta x$, for an arbitrary $\Delta x \neq 0$;
(c) The instantaneous rate of change of $f(x)$ at $x = 3$.

Solution. (a) For the first of the two average rates of change, we set $a = 3$ and $\Delta x = 0.1$. Then

$$\frac{\Delta y}{\Delta x} = \frac{f(a + \Delta x) - f(a)}{\Delta x} = \frac{f(3.1) - f(3)}{0.1} = \frac{3.1^2 - 3^2}{0.1} = \frac{9.61 - 9}{0.1} = \frac{0.61}{0.1} = 6.1.$$

Similarly, for $\Delta x = 0.01$, we have

$$\frac{\Delta y}{\Delta x} = \frac{f(3.01) - f(3)}{0.01} = \frac{3.01^2 - 3^2}{0.01} = \frac{9.0601 - 9}{0.01} = \frac{0.0601}{0.01} = 6.01.$$

(b) Here, we find that

$$\frac{\Delta y}{\Delta x} = \frac{f(a + \Delta x) - f(a)}{\Delta x} = \frac{f(3 + \Delta x) - f(3)}{\Delta x} = \frac{(3 + \Delta x)^2 - 3^2}{\Delta x}$$

$$= \frac{9 + 6\Delta x + (\Delta x)^2 - 9}{\Delta x} = \frac{6\Delta x + (\Delta x)^2}{\Delta x} = \frac{\Delta x(6 + \Delta x)}{\Delta x} = 6 + \Delta x. \qquad (2.3)$$

(c) It's quite clear that the the limit, as Δx approaches zero, of the right-hand side of (2.3) equals 6. But the left-hand and right-hand sides of (2.3) are equal, so the limit of the right-hand side must equal the limit of the left-hand side. And the limit of the left-hand side *is*, by Definition 2.1.2, the instantaneous rate of change of $f(x)$ at $x = 3$, also denoted $f'(3)$. In sum,

$$f'(3) = \lim_{\Delta x \to 0} \frac{\Delta y}{\Delta x} = \lim_{\Delta x \to 0} (6 + \Delta x) = 6.$$

Note that the evaluation of $f'(3)$, in part (c) of the above example, relied heavily on the *algebra* employed in part (b). Specifically, in part (b) we were able to simplify the *numerator* Δy of our difference quotient, to the point where we could factor Δx out of this numerator. We then *cancelled* this factor against the Δx in the denominator. This was crucial because, had a factor of Δx *remained* in the denominator, then letting $\Delta x \to 0$ in part (c) would have effectively left us with a *zero* in the denominator, and we know that can't be good!

Derivative computations will typically entail some type of "cancellation in numerator and denominator." However, that cancellation can take a variety of forms, one of which is illustrated in the following example.

Example 2.1.3 Let $h(x) = \sin(x)$. Find $h'(0)$. Use the "trigonometric limit formula"

$$\lim_{t \to 0} \frac{\sin(t)}{t} = 1. \tag{2.4}$$

(See Exercise 14, Sect. 1.7.4, above. Heuristically, this limit formula says that, as an angle shrinks to zero, its sine and its radian measure become very close to each other.)

Solution. By Definition 2.1.2 of the derivative, we have

$$h'(0) = \lim_{\Delta x \to 0} \frac{h(0 + \Delta x) - h(0)}{\Delta x} = \lim_{\Delta x \to 0} \frac{\sin(\Delta x) - \sin(0)}{\Delta x} = \lim_{\Delta x \to 0} \frac{\sin(\Delta x) - 0}{\Delta x}$$

$$= \lim_{\Delta x \to 0} \frac{\sin(\Delta x)}{\Delta x} = 1,$$

the last step by (2.4) (with $t = \Delta x$).

We will explore other derivative computations in the exercises below. In particular, we'll use the definition of the derivative to confirm our intuition about instantaneous rates of change in the water density context (Example 2.1.1) above.

We now wish to interpret the derivative geometrically. To do this, let's note that the average rate of change

$$\frac{\Delta y}{\Delta x} = \frac{f(a + \Delta x) - f(a)}{\Delta x}$$

is just the slope of the line through the points $(a, f(a))$ and $(a + \Delta x, f(a + \Delta x))$ on the graph of f. This line is called a *secant line* to the graph of $f(x)$, meaning a line that intersects this graph in (at least) two points (Fig. 2.1).

What does this have to do with derivatives? Well: note that, by the definition of the derivative and by the above geometric interpretation of $\Delta y/\Delta x$,

Fig. 2.1 An average rate of change is the slope of a secant line

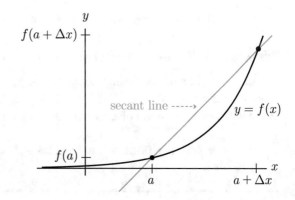

Fig. 2.2 As Δx shrinks, the secant lines become the tangent line

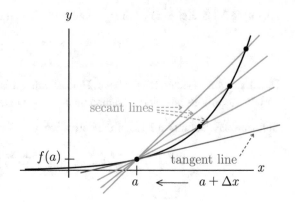

$$f'(a) = \lim_{\Delta x \to 0} \frac{\Delta y}{\Delta x} = \lim_{\Delta x \to 0} \frac{f(a + \Delta x) - f(a)}{\Delta x}$$

$$= \lim_{\Delta x \to 0} [\text{slope of the secant line through } (a, f(a)) \text{ and } (a + \Delta x, f(a + \Delta x))]$$

$$= \text{slope of the \textbf{tangent line} to the graph of } f(x) \text{ at the point } x = a, \qquad (2.5)$$

since, as $\Delta x \to 0$, the secant lines in question approach this tangent line. See Fig. 2.2. (The tangent line is the line "just touching" the graph of $f(x)$ at the point in question. Intuitively, one can think of the tangent line as the "secant line through $(a, f(a))$ and $(a + \Delta x, f(a + \Delta x))$, where Δx is infinitesimally small.")

We summarize:

> **The instantaneous rate of change, or derivative, $f'(a)$, equals the slope of the line tangent to the graph of $y = f(x)$ at the point $x = a$.**

The derivative at a point is the slope of the tangent line at that point

Example 2.1.4 Find the equation of the line tangent to $f(x) = x^2$ at $x = 3$.

Solution. This line passes through the point $(3, f(3)) = (3, 3^3) = (3, 9)$, and has slope equal to the derivative $f'(3)$ of $f(x)$ at $x = 3$. We saw in Example 2.1.2 that $f'(3) = 6$. So, by the initial-value formula (1.8), the tangent line has equation

$$y = f(3) + f'(3)(x - 3) = 3^2 + 6(x - 3) = 9 + 6x - 18 = 6x - 9.$$

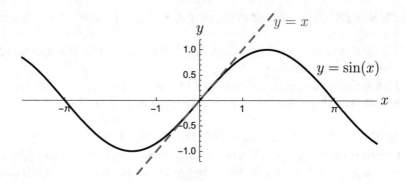

Fig. 2.3 The graph of $y = \sin(x)$ and its tangent line at $x = 0$

The above example illustrates a general fact. Suppose we have an arbitrary function $f(x)$, and an arbitrary point $x = a$, and suppose $f'(a)$ exists. Then the line tangent to the graph of f has slope $f'(a)$, and passes through the point $(a, f(a))$. Consequently, by the initial-value formula (1.8), this tangent line has equation

$$y = f(a) + f'(a)(x - a)$$ (2.6)

Equation of the line tangent to $y = f(x)$ **at** $x = a$

Again, this formula makes sense only in situations where $f'(a)$ exists.

For instance, Example 2.1.3 above tells us that the tangent line to the graph of $h(0) = \sin(x)$ at $x = 0$ (Fig. 2.3) has equation

$$y = h(0) + h'(0)(x - 0) = 0 + 1 \cdot x = x.$$

To summarize the geometry of rates of change: an average rate of change is the slope of a *secant* line; an instantaneous rate of change is the slope of a *tangent* line.

We will often refer to "the slope of $y = f(x)$ at $x = a$" when we mean "the slope of the line *tangent* to $y = f(x)$ at $x = a$." Again, this slope is just $f'(a)$ (when $f'(a)$ exists). So we think of the derivative of a function, at a given point, as telling us the slope of that function at that point.

2.1.1   Exercises

1. Let $f(x) = 2x^2 - 3$.

    (a) Find the average rate of change $\Delta y/\Delta x$ of $f(x)$ with respect to x, from $x = 2$ to $x = 2 + \Delta x$, for each of the following three values of Δx: $\Delta x = 0.1$, $\Delta x = 0.01$, $\Delta x = 0.001$.

(b) Based on part (a) above, what might you guess $f'(2)$ is equal to? Please explain your answer.

(c) Use algebra to show that the average rate of change of $f(x)$ with respect to x, from $x = 2$ to $x = 2 + \Delta x$, is $8 + 2\Delta x$.

(d) Find the instantaneous rate of change of $f(x)$ at $x = 2$.

(e) Find the equation of the line tangent to the graph of $f(x)$ at $x = 2$.

2. Repeat Exercise 1 above with the same function $f(x)$, but this time, at $x = -1$. (That is: for part (a) of the present exercise, compute average rates of change of $f(x)$ from $x = -1$ to $x = -1 + \Delta x$, for the same three values of Δx as in Exercise 1(a). And so on.)

3. Let $g(x) = -x^3 + 1$.

(a) Show that the average rate of change of $g(x)$ with respect to x, from $x = 4$ to $x = 4 + \Delta x$, is $-48 - 12\Delta x - (\Delta x)^2$. Hint: $-(4 + \Delta x)^3 = -64 - 48\Delta x - 12(\Delta x)^2 - (\Delta x)^3$.

(b) Find $g'(4)$.

4. Let m and b be constants, and let $y = f(x) = mx + b$.

(a) Recalling that the derivative of a function at a point measures the slope of that function at that point, determine, without any computation, what $f'(a)$ should be for any real number a. (Your answer will involve one or more of the constants m and b.) Please explain your answer.

(b) Verify your answer from part (a) of this exercise using the definition of the derivative. Specifically:

    (i) Compute the average rate of change of $y = f(x)$ from $x = a$ to $x = a + \Delta x$, where a is any real number and Δx is any nonzero number.

    (ii) Use your answer from part (i) to evaluate $f'(a)$.

(c) True or false: for a linear function, average and instantaneous rates of change are always equal. Please explain your answer.

5. Let $D(C)$ be as in Example 2.1.1 above.

(a) Show that the average rate of change of D with respect to C, over the interval $[1, 1 + \Delta x]$, is $0.04896 - 0.008\Delta x$.

(b) Use part (a) to find $D'(1)$. Does this result agree with our conclusion concerning "the instantaneous rate of change of D with respect to C, at $C = 1$," from part (b) of Example 2.1.1 above?

(c) Show that the average rate of change of D with respect to C, over the interval $[2, 2 + \Delta x]$, is $0.03296 - 0.008\Delta x$.

(d) Use part (c) to find $D'(2)$. Does this result agree, to several decimal places, with part (b) of Example 2.1.1 (in the case $C = 2$)?

6. Let $g(x) = \cos(x)$.

(a) Show that the average rate of change of $g(x)$, from $x = \pi/2$ to $x = \pi/2 + \Delta x$, is $-\sin(\Delta x)/\Delta x$. Hint: use the trigonometric identity $\cos(\pi/2 + \theta) = -\sin(\theta)$.

(b) Use the definition of the derivative to find $g'(\pi/2)$. Hint: use the "trigonometric limit" (2.4) above.

7. Let $f(x) = \sqrt{x}$.

(a) Show that the average rate of change of $f(x)$ with respect to x, from $x = 64$ to $x = 64 + \Delta x$, is

$$\frac{\sqrt{64 + \Delta x} - 8}{\Delta x}.$$

(b) Multiply the numerator and denominator of your answer from part (a) by $\sqrt{64 + \Delta x} + 8$, and then do some algebra to simplify, to show that the average rate of change from part (a) equals

$$\frac{1}{\sqrt{64 + \Delta x} + 8}.$$

(c) Use your result from part (b) to show that $f'(64) = 1/16$.

2.2   Local Linearity (Differentiability)

We begin with the following.

Definition 2.2.1 We say that the function $y = f(x)$ is **locally linear**, or **differentiable**, at the point $x = a$ if the limit

$$\lim_{\Delta x \to 0} \frac{f(a + \Delta x) - f(a)}{\Delta x} \tag{2.7}$$

exists.

We simply say "f is locally linear" (or "differentiable") if it's locally linear at *all* points in its domain.

We saw in the previous section that the limit (2.7), if it exists, is the derivative $f'(a)$ of $y = f(x)$ at $x = a$. So: "$y = f(x)$ is locally linear, or differentiable, at the point $x = a$" simply means "the derivative $f'(a)$ exists."

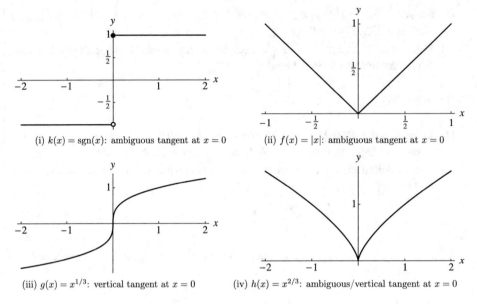

(i) $k(x) = \text{sgn}(x)$: ambiguous tangent at $x = 0$ (ii) $f(x) = |x|$: ambiguous tangent at $x = 0$

(iii) $g(x) = x^{1/3}$: vertical tangent at $x = 0$ (iv) $h(x) = x^{2/3}$: ambiguous/vertical tangent at $x = 0$

Fig. 2.4 Some functions with curious behavior at $x = 0$

Geometrically, existence of $f'(a)$ means existence of the slope of the line tangent to $y = f(x)$ at $x = a$. To understand what it means for this slope to exist, it is instructive to first consider some situations where it *does not* exist.

Example 2.2.1 Consider the functions (Fig. 2.4)

$$k(x) = \text{sgn}(x) = \begin{cases} -1 & \text{if } x < 0, \\ 1 & \text{if } x \geq 0, \end{cases} \quad f(x) = |x|, \quad g(x) = \sqrt[3]{x} = x^{1/3}, \quad h(x) = \sqrt[3]{x^2} = x^{2/3}.$$

Explain geometrically why each of these functions fails to be differentiable at $x = 0$.

Solution. Because of the "jump" in the graph of $k(x) = \text{sgn}(x)$ at $x = 0$, there's no unambiguous way to define a line "just touching" this graph at $x = 0$. That is, there's no unambiguous definition of a tangent line, and therefore no unambiguous definition of a derivative, at this point.

Regarding $f(x) = |x|$: we might imagine that *many* lines are just touching the graph of this function at $x = 0$. For example, the lines $y = -x/2$, $y = -x/3$, and $y = 3x/4$ all "hinge" on this graph at this point. These lines all have different slopes, so there's no unambiguous definition of $f'(0)$.

The graph of $g(x) = x^{1/3}$ *does* have a unique tangent line at $x = 0$, but this tangent line is vertical, and therefore does not have a finite slope. Therefore, we can not define $g'(0)$.

Finally: as with $y = |x|$, one can imagine many different lines "hinging" on the graph of $h(x) = x^{2/3}$ at $x = 0$. One could even imagine a vertical line balancing on this graph there. So this function is about as bad as one can get, from the point of view of differentiability: not only are there too many possible candidates for $h'(0)$, but one of those candidates is undefined.

We have some sense, then, of how a derivative can go wrong. But now, let's reflect on what it means for a derivative to go *right*. What we wish to argue is that:

> **If $f'(a)$ exists then, in a vicinity of $x = a$,**
> **$f(x)$ looks a lot like its tangent line at that point.**

To make this argument, we observe the following. Again, $f'(a)$ is an instantaneous rate of change, while the difference quotient

$$\frac{f(a + \Delta x) - f(a)}{\Delta x}$$

is an average rate of change. And this average rate of change should be pretty close to this instantaneous rate of change if we average over a short enough interval. That is, geometrically speaking: the slope of the secant line should be approximately equal to the slope of the tangent line, if the latter slope exists and if the two points on the secant line are close enough to each other. In other words, we should expect that, if $f'(a)$ exists, then

$$\frac{f(a + \Delta x) - f(a)}{\Delta x} \approx f'(a) \quad \text{for } \Delta x \text{ small enough.} \tag{2.8}$$

If we multiply both sides of Eq. (2.8) by Δx, and then add $f(a)$ to both sides, we get

$$f(a + \Delta x) \approx f(a) + f'(a)\Delta x \quad \text{for } \Delta x \text{ small enough.} \tag{2.9}$$

Next, we give a new name to $a + \Delta x$. Let's simply call it x. Then $\Delta x = x - a$, so Eq. (2.9) says

$$f(x) \approx f(a) + f'(a)(x - a) \quad \text{for } x - a \text{ small enough.} \tag{2.10}$$

The right-hand side of (2.10) is, as we saw in the previous section, just the tangent line to $y = f(x)$ at $x = a$. (See Eq. (2.6).) So (2.10) tells us that $f(x)$ *and the tangent line to the graph of $f(x)$ at $x = a$ are approximately the same*, if x is close to a. This is exactly what we wanted to demonstrate. And it justifies the term "locally linear:" if $f'(a)$ exists then, at least locally (near $x = a$), the graph of $f(x)$ looks like a line. (And not just any line; it looks like its tangent line at this point.)

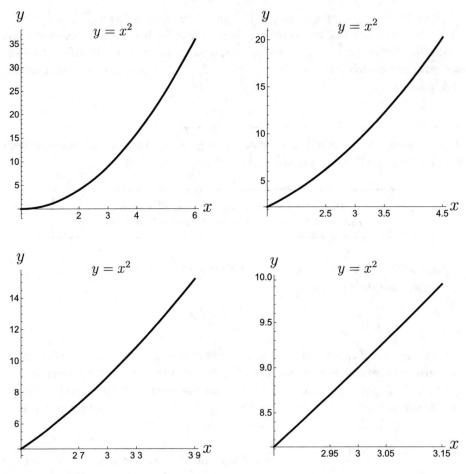

Fig. 2.5 Zooming in on $f(x) = x^2$ near $x = 3$

For example, Fig. 2.5 shows graphs of $f(x) = x^2$ at successively finer scales, all zeroing in on the point $x = 3$. (Compare with Examples 2.1.2 and 2.1.4 above.) Note that the axes do not intersect at $(0, 0)$ in any of these graphs except the first.

Remark 2.2.1 The existence of $f'(a)$ does not tell us how *far* we need to zoom in, before $f(x)$ "looks linear" near $x = a$. It merely tells us that it will do so at *some* "magnification." This is to be contrasted with functions that fail to be differentiable at a given point; such functions will not look linear no matter how far we zoom in on that point. See the exercises below.

The intuition behind all of the above examples is that a function is differentiable at a point if it has no jumps, breaks, vertical tangent lines, or sharp corners at that point. For example, each of the functions of Example 2.2.1 above is differentiable at every number x *except* $x = 0$. It may be shown that polynomials are differentiable everywhere, as are the functions $y = \cos(x)$ and $y = \sin(x)$. And other trigonometric functions, such as $y = \tan(x)$, are differentiable except at points where they are undefined. (For example, $y = \cot(x)$ is nondifferentiable precisely at the integer multiples of π, since $\cot(x) = \cos(x)/\sin(x)$, and $\sin(x)$ is zero precisely when $x = k\pi$ for some integer k.) Similarly, *rational functions*, meaning functions that are defined as quotients of polynomials, e.g.

$$q(x) = \frac{x^4 + 1}{x^2 - 2x - 3} \quad \text{or} \quad m(z) = \frac{45z^{46} - 26z^{15} + 3z - 4}{14z^{100} - 100z^{14} + 1401},$$

are differentiable everywhere except at points where the denominator equals zero.

2.2.1 Continuity

We say that a function f is **continuous at a point** $x = a$ if

- it is defined at the point $x = a$, and
- we can make $f(x)$ as close as we want to $f(a)$ if we take x to be close enough to a.

This second condition can also be expressed in the following form (due, in essence, to Augustin Cauchy in the early 1800's):

> Given any positive number ϵ (the Greek letter "epsilon;" think of ϵ as "how close we want to make $f(x)$ to $f(a)$"), there is always a positive number δ (the Greek letter "delta;" think of δ as "how close we need to make x to a"), such that whenever the change $|x - a|$ in the input is less than δ, then the corresponding change $|f(x) - f(a)|$ in the output will be less than ϵ.

A function is said to be continuous on a set of real numbers if it is continuous at each point of the set.

Cauchy's definition of continuity is quite technical. For our purposes, it will suffice to think of continuity in a more intuitive way: a function is continuous on an interval if the graph of the function has no gaps or jumps in it, on that interval. That is, a continuous function is one that you can draw without picking up your pencil.

It's not hard to see that differentiability is *stronger* than continuity. That is, if f is differentiable at $x = a$, then it must be continuous there. The converse is false: continuity need not imply differentiability. For instance, the functions f, g, and h of Example 2.2.1 are all continuous at $x = 0$, but none are differentiable there. (The function k of Example 2.2.1 is neither continuous nor differentiable at $x = 0$.)

There are examples of functions that are continuous but nondifferentiable at every single real number x! But we won't encounter such functions in this text.

2.2.2 Exercises

1. Using Sage, plot the function $f(x) = 3x^5 - 5x^3$ on each of the following domains (all of which are centered on $x = 2$): $[0, 4]$, $[1, 3]$, $[1.9, 2.1]$, $[1.99, 2.01]$.

   (a) Describe what you see (in terms of the apparent shape of $f(x)$, as you zoom in further and further).
   (b) What one-word, or two-word, adjective from Sect. 2.2 above describes this function $f(x)$ at $x = 2$?
   (c) From your plot over the interval $[1.99, 2.01]$, in part (a) above, choose two distinct points on the graph of $f(x)$. Label these points with their coordinates. (Read the coordinates off of the axes as well as you can. Or plug the x-values for your chosen points into the formula for $f(x)$, to get the corresponding y-values.)
   (d) Compute and write down the slope of the line between these two points. Use this result to estimate $f'(2)$.

2. Repeat Exercise 1 above for the function $f(x) = 8\sqrt{x + 2}$. Use the same four intervals here as you did in Exercise 1. Again, use your result to estimate $f'(2)$.
3. Repeat Exercise 1 above for the function $f(x) = 4/x$. Use the same four intervals here as you did in Exercise 1. Again, use your result to estimate $f'(2)$.
4. Using Sage or similar computer software, plot the function $g(x) = x^2 - 10|x - 2|$ on each of the domains $[0, 4]$, $[1.9, 2.1]$, $[1.99, 2.01]$, and $[1.999, 2.001]$. (In Sage, you can use abs() to take absolute values.)

   (a) Describe what you see (in terms of the apparent shape of $g(x)$, as you zoom in further and further).
   (b) Does the graph of $g(x)$ look more and more like that of a line, as you zoom in more and more? Is $g(x)$ locally linear at $x = 2$, or not?

5. Repeat Exercise 4 above for the function $g(x) = x^2 - 20(x - 2)^{4/5}$. (You should input this function to Sage as x^2-20*((x-2)^4)^(1/5). Using ((x-2)^4)^(1/5) instead of (x-2)^(4/5) will avoid the issue of raising negative numbers to fractional exponents.)
6. At how many points, and at which points, on the interval $[0, 3]$ does the function $q(x) = 1 - 2|\sin(2\pi x)|$ fail to be locally linear? Answer by using Sage to sketch this function. (Recall that, in Sage, you can use sin() for the sine function.)

7. Note: answers to the following exercise will vary widely. Use your intuition; you don't need to know any formulas from physics.

Imagine dropping a ball from a height of four feet.

(a) Let $H(t)$ be the height of this ball at time t, with H in feet above the ground, and t in seconds from the instant the ball is dropped. Draw, by hand, a *rough sketch* of $H(t)$. Sketch over a large enough domain to capture several bounces of the ball.

Don't worry about being "to scale." (For example, you don't need specific equations that would tell you how long it would take for the ball to reach the ground.) Just try to sketch the general shape of the graph.

(b) If you were to zoom in on your graph from part (a), near the instant where the ball first hits the ground, what do you think you'd see? Draw a very rough sketch of this.

(c) Is the function that you sketched in part (a) locally linear at all points in your chosen domain? If not, at which points does it fail to be locally linear?

8. As with the previous exercise, answers to this one will vary widely. Again, focus on intuition and general ideas, and don't worry about specific numbers.

Draw a very rough sketch of the *velocity* $H'(t)$ of the ball in Exercise 7, for all values of t that you included in your graph in Exercise 7. **Important note:** if the ball is falling, then its velocity is negative (because its height above the ground is *decreasing*). If the ball is rising, then its velocity is positive.

Is $H'(t)$ continuous at all points in its domain? If not, then when does it fail to be continuous? (You don't need to give specific values of t; just describe the ball's position at points where its velocity is discontinuous.)

2.3 Formulas and Rules for Derivatives

The process of finding derivatives of functions is called **differentiation.**

Given a formula for a function f we can, at least in theory, differentiate f, at any point x where f is locally linear. Namely, we can apply the formula

$$f'(x) = \lim_{\Delta x \to 0} \frac{f(x + \Delta x) - f(x)}{\Delta x}. \tag{2.11}$$

This formula represents the same definition of the derivative as was given in Sect. 2.1. But in that section, we wrote "a" to denote our input to f'. Here, we are instead using an "x" for our input to f', to emphasize the fact that this input is a *variable*. That is, we now are thinking globally—we are thinking of $f'(x)$ as a *function* of x, just as $f(x)$ is. Specifically: $f'(x)$, at any point x in its domain, equals the slope of $f(x)$ at that point. (Note that the domain of f' might be smaller than that of f, since there may be points at which $f(x)$ is defined, but not locally linear.)

Derivative calculations using Eq. (2.11) can sometimes be tedious, as you may have noticed in the course of working examples and exercises from Sect. 2.1. Our goal over the next few sections is to expedite the differentiation process, using two types of tools. The first variety of tool is the *differentiation formula*. By this, we mean an equation that tells us how to differentiate a particular function, or family of functions. The idea here is that certain kinds of functions—power functions, sines, and cosines, for example—arise frequently. If we catalog the derivatives of such functions, we can then simply "look up" (or remember) those derivatives when we need them.

The second type of tool is the *differentiation rule*. By this we mean a general prescription for expressing the derivative of a complex function in terms of the derivatives of simpler "building blocks."

The idea here is that many complex functions are built up from simpler ones. For instance, the function given by the expression

$$3x^7 + 8\sin(x)\sqrt{x} \tag{2.12}$$

is built up from the basic functions x^7, $\sin(x)$, and $\sqrt{x}$. In fact, since $\sqrt{x} = x^{1/2}$, we can think of x^7 and $\sqrt{x}$ as two different instances of the general "power function" x^p.

So, if we have differentiation *formulas* that tell us how to differentiate basic things like x^p and $\sin(x)$, together with differentiation *rules* that tell us how to express the derivative of a complex function in terms of simpler constituent parts, then we *will* be able to find derivatives of complicated things like (2.12).

2.3.1 Differentiation Formulas

Constant functions. Consider the constant function $f(x) = c$, where c is a fixed real number. The graph of this function is a horizontal line (of height c), and such a line has slope 0 at all points x. But the derivative *is* the slope, so we conclude that

> **The derivative $f'(x)$ of a constant function $f(x) = c$ is zero at all points x.**

The constant function formula

Power functions. Before proceeding further, we introduce some new notation. Namely, suppose $y = f(x)$ is a function of x. We will sometimes write

$$\frac{d}{dx}[f(x)],$$

pronounced "dee dee x of f of x," to denote the derivative $f'(x)$. This new notation has the advantage of allowing us to refer to the derivative of a function given by a formula without giving that function an explicit name. For example, the derivative of the function defined by (2.12) may be expressed as

$$\frac{d}{dx}\left[3x^7 + 8\sin(x)\sqrt{x}\right].$$

And the above statement "The derivative $f'(x)$ of a constant function $f(x) = c$ is zero at all points x" may simply be written

$$\frac{d}{dx}[c] = 0 \text{ for any constant } c.$$

(In writing a derivative formula using this new notation, we will implicitly assume that the formula holds at all points where the function to be differentiated is locally linear.) Of course we are not restricted to x as a variable name; we might consider

$$\frac{d}{dz}\left[\frac{z^3 + 4^z}{1 + \cos(z)}\right],$$

for example.

With this notation in hand, we now ask: what is

$$\frac{d}{dx}[x^p]$$

for an arbitrary real number p? That is, what is the derivative of a *power function*? We first consider some examples.

Example 2.3.1 Find the derivative $\dfrac{d}{dx}[x^p]$ for each of the following values of p: (a) $p = 1$; (b) $p = 2$; (c) $p = -1$.

Solution. (a) The function $f(x) = x^1 = x$ gives a line with slope 1, so $\dfrac{d}{dx}[x^1] = 1$.

(b) We compute, using formula (2.11), that

$$\frac{d}{dx}[x^2] = \lim_{\Delta x \to 0} \frac{(x + \Delta x)^2 - x^2}{\Delta x} = \lim_{\Delta x \to 0} \frac{x^2 + 2x\Delta x + (\Delta x)^2 - x^2}{\Delta x}$$

$$= \lim_{\Delta x \to 0} \frac{2x\Delta x + (\Delta x)^2}{\Delta x} = \lim_{\Delta x \to 0} \frac{\Delta x(2x + \Delta x)}{\Delta x} = \lim_{\Delta x \to 0} (2x + \Delta x) = 2x.$$

(You should compare both the computations and the result here to those of Example 2.1.2, which investigated this same derivative, but specifically at the point $x = 3$.)

(c) We have

$$\frac{d}{dx}\left[x^{-1}\right] = \frac{d}{dx}\left[\frac{1}{x}\right] = \lim_{\Delta x \to 0} \frac{1}{\Delta x}\left(\frac{1}{x + \Delta x} - \frac{1}{x}\right) = \lim_{\Delta x \to 0} \frac{1}{\Delta x}\left(\frac{x - (x + \Delta x)}{(x + \Delta x)x}\right)$$

$$= \lim_{\Delta x \to 0} \frac{1}{\Delta x}\left(\frac{-\Delta x}{(x + \Delta x)x}\right) = \lim_{\Delta x \to 0} \frac{-1}{(x + \Delta x)x} = \frac{-1}{x^2} = -x^{-2}.$$

(To get the third equality, we found a common denominator for the two fractions inside the large parentheses.)

The above example suggests a pattern, which we encapsulate as follows:

$$\boxed{\frac{d}{dx}[x^p] = px^{p-1} \text{ for any real number } p}$$

The power formula

(The power formula is also sometimes called the power *rule*, but we are reserving the term "rule" for a different context; see Sect. 2.3.2, below.)

There is one function to which both the constant and the power formula apply: the function $f(x) = 1 = x^0$. Note that the constant formula tells us that, for this function, $f'(x) = 0$, while the power formula tells us that $f'(x) = \frac{d}{dx}[x^0] = 0x^{0-1} = 0$. That is, the two formulas give the same answer in this case, as they had better! (Technically, there's a problem with defining $0x^{0-1}$ if $x = 0$. But a direct calculation, using the definition of the derivative, shows that $f'(0) = 0$.)

We have by no means proved the power formula for all possible powers p. Note that there are many types of exponents to consider: positive and negative integers; positive and negative rational numbers (that is, fractions), *irrational* powers like $\sqrt{2}$ and π, and so on. Other references will supply proofs for some or all of these kinds of exponents. We will content ourselves, here, with simply *stating* the formula, and reassuring ourselves with the few special cases where we've actually done the computations. And if you are still skeptical, the exercises below explore a few more special cases, for additional reassurance.

In the following example, we investigate another family of functions, where the variable is in the *exponent*, rather than being in the *base*, as it is for the power functions above.

Example 2.3.2 (*Exponential functions*) Let b be a positive constant. Show that

$$\boxed{\frac{d}{dx}[b^x] = \ln(b)b^x} \tag{2.13}$$

Derivative of an exponential function

where $\ln(b)$, called **the natural logarithm of b**, is defined by the formula

$$\ln(b) = \lim_{\Delta x \to 0} \frac{b^{\Delta x} - 1}{\Delta x}. \tag{2.14}$$

Solution. Using the fact that $b^{x+y} = b^x b^y$ for all real numbers x and y, we find that

$$\frac{d}{dx}[b^x] = \lim_{\Delta x \to 0} \frac{b^{x+\Delta x} - b^x}{\Delta x} = \lim_{\Delta x \to 0} \frac{b^x b^{\Delta x} - b^x}{\Delta x} = \lim_{\Delta x \to 0} \frac{b^x (b^{\Delta x} - 1)}{\Delta x}. \tag{2.15}$$

Now note that the factor b^x, on the right-hand side of Eq. (2.15), is independent of Δx. So we may, in fact, move this factor out in front of the "$\lim_{\Delta x \to 0}$." We do so using the eminently plausible, and *true*, fact (whose proof we omit) that "the limit of a constant multiple equals the constant multiple of the limit." That is, if $Q(\Delta x)$ is some quantity that approaches a number L as $\Delta x \to 0$, and if c is a number that does not depend on Δx, then $c\, Q(\Delta x)$ approaches cL as $\Delta x \to 0$. This fact, together with (2.15), tell us that

$$\frac{d}{dx}[b^x] = b^x \lim_{\Delta x \to 0} \frac{b^{\Delta x} - 1}{\Delta x} = \ln(b)b^x,$$

as required.

A few comments on Example 2.3.2 are in order. First: we require that $b > 0$ to avoid difficulties like square roots of negative numbers. For example, were we to take $b = -1$, then letting $x = 1/2$ would give $b^x = (-1)^{1/2} = \sqrt{-1}$, which is not defined (as a real number). Stipulating that $b > 0$ eliminates such issues.

Second: $\ln(b)$ depends on b, but not on x. So the above example demonstrates the fundamental fact that *the derivative of an exponential function is a constant (with respect to the independent variable) times that function.* We've encountered functions that behave like this before—see the subsection "Functions proportional to their rates of change," in Sect. 1.7, above. There, we considered quantities P satisfying rate equations of the form $P' = kP$. In light of Example 2.3.2, then, we should expect that such quantities P are exponential in nature. This expectation will be borne out in Sect. 3.1 below.

Third: the definition (2.14) of $\ln(b)$ is a bit complicated. And in general, there's no algebra we can do to extract precise numerical values from this definition—except for the case $b = 1$, since

$$\ln(1) = \lim_{\Delta x \to 0} \frac{1^{\Delta x} - 1}{\Delta x} = \lim_{\Delta x \to 0} \frac{1 - 1}{\Delta x} = \lim_{\Delta x \to 0} \frac{0}{\Delta x} = \lim_{\Delta x \to 0} 0 = 0.$$

For more general $b > 0$, we can nonetheless obtain, from (2.14), an arbitrarily good *approximation* to $\ln(b)$. We do so by evaluating the difference quotient in (2.14) at a suitably small value of Δx. For example: putting $b = 2$ and $\Delta x = 0.000001$ gives

$$\ln(2) \approx \frac{2^{0.000001} - 1}{0.000001} = 0.69314\ldots,$$

which is correct to five decimal places.

Fourth: you may have encountered natural logarithms in other contexts, where you probably saw them defined differently. We'll see in Chap. 3 that other, perhaps more familiar, definitions of $\ln(b)$ are equivalent to the one given above.

Fifth, and last: note the fundamental difference between a *power function*, where the input, or independent variable, is in the base, and an *exponential function*, where the input appears in the exponent. When you differentiate a power function, the original exponent drops by 1, and also appears as a factor in your derivative. When you differentiate an exponential function, the exponent doesn't change, but the natural logarithm of the *base* appears as a factor. For example,

$$\frac{d}{dx}[x^4] = 4x^3 \quad \text{but} \quad \frac{d}{dx}[4^x] = \ln(4)4^x;$$
$$\frac{d}{dx}[x^\pi] = \pi x^{\pi-1} \quad \text{but} \quad \frac{d}{dx}[\pi^x] = \ln(\pi)\pi^x.$$

Remark on notation. As with trigonometric functions, we sometimes use shorthand to write expressions involving the natural logarithm function. Specifically: we sometimes write $\ln b$ for $\ln(b)$; we sometimes write $\ln^k(b)$, or even $\ln^k b$, for $(\ln(b))^k$.

The final general category of function that we will consider, in this section, is the category of trigonometric functions. For example, let $h'(x) = \sin(x)$: we saw in Example 2.1.3 above that $h'(0) = 1$. Using ideas from that example, together with some trigonometric identities, we can similarly show that $h'(x) = \cos(x)$ for all x. And we can analogously compute $\frac{d}{dx}[\cos(x)]$ and $\frac{d}{dx}[\tan(x)]$. See the Exercises below. We compile these results, along with the others catalogued above, into Table 2.1, below.

Here c and p can be any real numbers, and b can be any positive number. Also, recall that $\sec(x) = 1/\cos(x)$, and that $\sec^2(x)$ is shorthand for $(\sec(x))^2$.

Again, remember that the input to the trigonometric functions is always measured in radians; the above formulas are not correct if x is measured in degrees. There are similar formulas if you insist on using degrees, but they are more complicated. This is the principal reason we work in radians—the derivative formulas are nice!

2.3.2 Differentiation Rules

Since basic functions are combined in various ways to make more complicated ones, we need to know how to differentiate *combinations*. For example, suppose we add the functions $g(x)$ and $h(x)$, to get $f(x) = g(x) + h(x)$. It may be shown, then, that f is differentiable too, and moreover, that "the rate at which f changes is the sum of the separate rates at which

g and h change," or "the derivative of the sum is the sum of the derivatives." More formally, we have

$$\frac{d}{dx}[f(x) + g(x)] = f'(x) + g'(x)$$

The sum rule

For example,

$$\text{if } f(x) = \tan(x) + x^{-6}, \text{ then } f'(x) = \sec^2(x) - 6x^{-7};$$

$$\frac{d}{dw}\left[2^w + \cos(w)\right] = \ln(2)2^w - \sin(w).$$

Likewise, if we multiply any differentiable function f by a constant c, then, as one can show, the product $g(x) = cf(x)$ is also differentiable, and moreover "the derivative of the constant multiple equals the constant multiple of the derivative," or "rescaling a function vertically rescales its rate of change by the same factor." That is,

$$\frac{d}{dx}[cf(x)] = cf'(x)$$

The constant multiple rule

For example, $\dfrac{d}{dx}[5\sin(x)] = 5\cos(x)$. Likewise, if $g(z) = 25z^3$, then $g'(z) = 25 \cdot 3z^2 = 75z^2$. However, the rule does *not* tell us how to find the derivative of $\sin(5x)$, because $\sin(5x) \neq 5\sin(x)$. To work this one out, we will need the **chain rule**, which describes how to differentiate compositions $f(g(x))$, in terms of the derivatives f' and g'. See the following section. Subsequent sections will also investigate derivatives of products $f(x)g(x)$ and quotients $f(x)/g(x)$.

With just the few facts already laid out, we can differentiate a variety of functions given by formulas. Here are a couple more simple examples.

Example 2.3.3 Find:

(a) $\dfrac{d}{dq}\left[-\dfrac{7\cos(q)}{13} + 5\sqrt[3]{q^{11}} + \dfrac{1}{\sqrt{q}} + \dfrac{q^5}{4923} + 4923\,5^q - 23944923\pi^\pi\right].$

(b) The derivative of the general **polynomial** function

$$P(x) = a_n x^n + a_{n-1} x^{n-1} + \cdots + a_2 x^2 + a_1 x + a_0.$$

Here $a_n, a_{n-1}, \ldots, a_2, a_1, a_0$ are all constants, and n is a positive integer, called the **degree** of the polynomial.

Table 2.1 A short table of derivative formulas

Function $f(x)$	Derivative $f'(x) = \dfrac{d}{dx}[f(x)]$
c	0
x^p	px^{p-1}
b^x	$\ln(b)b^x$
$\sin(x)$	$\cos(x)$
$\cos(x)$	$-\sin(x)$
$\tan(x)$	$\sec^2(x)$

Solution. (a) First, we express the second and third summands in terms of powers, so that we can use the power formula. Then we differentiate, using Table 2.1 and the sum and constant multiple rules, above. We get

$$\frac{d}{dq}\left[-\frac{7\cos(q)}{13}+5\sqrt[3]{q^{11}}+\frac{1}{\sqrt{q}}+\frac{q^5}{4923}+4923\,5^q-23944923\pi^\pi\right]$$

$$=\frac{d}{dq}\left[-\frac{7\cos(q)}{13}+5q^{11/3}+q^{-1/2}+\frac{1}{4923}q^5+4923\,5^q-23944923\pi^\pi\right]$$

$$=-\frac{7}{13}\frac{d}{dq}[\cos(q)]+5\frac{d}{dq}[q^{11/3}]+\frac{d}{dq}[q^{-1/2}]+\frac{1}{4923}\frac{d}{dq}[q^5]+4923\frac{d}{dq}[5^q]-23944923\frac{d}{dq}[\pi^\pi]$$

$$=-\frac{7}{13}(-\sin(q))+5\cdot\frac{11}{3}q^{8/3}-\frac{1}{2}q^{-3/2}+\frac{1}{4923}\cdot 5q^4+4923\cdot\ln(5)5^q-23944923\cdot 0$$

$$=\frac{7}{13}\sin(q)+\frac{55}{3}q^{8/3}-\frac{1}{2}q^{-3/2}+\frac{5q^4}{4923}+4923\ln(5)5^q.$$

Although not necessary, we can rewrite the final answer by expressing the fractional powers of q in terms of roots and integers powers; we get

$$\frac{d}{dq}\left[-\frac{7\cos(q)}{13}+5\sqrt[3]{q^{11}}+\frac{1}{\sqrt{q}}+\frac{q^5}{4923}+4923\,5^q-23944923\pi^\pi\right]$$

$$=\frac{7}{13}\sin(q)+\frac{55}{3}\sqrt[3]{q^8}-\frac{1}{2\sqrt{q^3}}+\frac{5q^4}{4923}+4923\ln(5)5^q.$$

(b) The polynomial $P(x)$ is a sum of terms, each of which is a constant multiple of an integer power of the input variable. (A polynomial of degree 1 is just a linear function.) The derivative of $P(x)$, by the sum and constant multiple rules and the power and constant formulas above, is given by

$$P'(x)=na_n x^{n-1}+(n-1)a_{n-1}x^{n-2}+\cdots+2a_2 x+a_1.$$

2.3.3   Exercises

2.3.3.1 Part 1: Differentiation Using the Definition of the Derivative

For each exercise in this section, use the definition (2.11) to deduce the indicated derivative formula.

1. (a) Show that, if $f(x) = \sqrt{x}$, then $f'(x) = \dfrac{1}{2\sqrt{x}}$. For this exercise, you might wish to use the algebra "trick" of multiplying both numerator and denominator of a fraction by the same thing. More specifically, note that

$$f'(x) = \lim_{\Delta x \to 0} \frac{\sqrt{x + \Delta x} - \sqrt{x}}{\Delta x} = \lim_{\Delta x \to 0} \frac{\sqrt{x + \Delta x} - \sqrt{x}}{\Delta x} \cdot \frac{\sqrt{x + \Delta x} + \sqrt{x}}{\sqrt{x + \Delta x} + \sqrt{x}}$$

$$= \lim_{\Delta x \to 0} \frac{(\sqrt{x + \Delta x} - \sqrt{x})(\sqrt{x + \Delta x} + \sqrt{x})}{\Delta x(\sqrt{x + \Delta x} + \sqrt{x})}.$$

Do the algebra in the numerator; you should end up with a numerator that involves no square roots. (Use the fact that $\left(\sqrt{Y}\right)^2 = Y$ for any Y.) You should then be able to cancel out a Δx in your numerator and denominator. Then compute what happens to what's left, as $\Delta x \to 0$.

   (b) Explain why part (a) of this exercise agrees with the case $p = 1/2$ of the power formula.

2. (a) Show that, if $g(x) = \dfrac{1}{\sqrt{x}}$, then $g'(x) = -\dfrac{1}{2\sqrt{x^3}}$. Hint: obtaining a common denominator gives

$$g'(x) = \lim_{\Delta x \to 0} \frac{1}{\Delta x}\left(\frac{1}{\sqrt{x + \Delta x}} - \frac{1}{\sqrt{x}}\right) = \lim_{\Delta x \to 0} \frac{1}{\Delta x}\left(\frac{\sqrt{x} - \sqrt{x + \Delta x}}{\sqrt{x(x + \Delta x)}}\right).$$

Multiply the right-hand side by

$$\frac{\sqrt{x} + \sqrt{x + \Delta x}}{\sqrt{x} + \sqrt{x + \Delta x}}$$

and simplify. You should then be able to cancel out a Δx top and bottom, and then take the appropriate limit.

   (b) Explain why part (a) of this exercise agrees with the case $p = -1/2$ of the power formula.

3. In this exercise we calculate the derivative of $f(x) = x^4$.

   (a) Expand $f(x + \Delta x) = (x + \Delta x)^4 = (x + \Delta x)(x + \Delta x)(x + \Delta x)(x + \Delta x)$ as a sum of 16 terms.

   (b) Group the terms in part (a) of this exercise so that $f(x + \Delta x)$ has the form

$$Ax^4 + B\Delta x + C(\Delta x)^2 + D(\Delta x)^3 + E(\Delta x)^4 ,$$

where there are no Δx's in the terms $A, B, C, D,$ or E.

(c) Compute the quotient $\dfrac{f(x + \Delta x) - f(x)}{\Delta x}$, taking advantage of part (b) of this exercise.

(d) Now find

$$\lim_{\Delta x \to 0} \frac{f(x + \Delta x) - f(x)}{\Delta x};$$

this is the derivative of x^4. Is your result here compatible with the power formula for the derivative of x^p ?

4. In this exercise we calculate the derivative of $f(x) = x^n$ when n is a negative integer, assuming that we know how to differentiate x^m for m a positive integer.
First write $n = -m$, so m is a positive integer. Then $f(x) = x^{-m} = 1/x^m$.

(a) Use the reciprocal rule (see Sect. 2.5.2) and this new expression for f to find $f'(x)$.
(b) Do some algebra to re-express $f'(x)$ as nx^{n-1}.

5. (a) Show that, if $k(x) = \dfrac{1}{x^2}$, then $k'(x) = -\dfrac{2}{x^3}$. Hint: obtaining a common denominator gives

$$k'(x) = \lim_{\Delta x \to 0} \frac{1}{\Delta x}\left(\frac{1}{(x + \Delta x)^2} - \frac{1}{x^2}\right) = \lim_{\Delta x \to 0} \frac{1}{\Delta x}\left(\frac{x^2 - (x + \Delta x)^2}{x^2(x + \Delta x)^2}\right).$$

Perform some algebra in the numerator; then cancel out a Δx top and bottom; the consider what happens as $\Delta x \to 0$.

(b) Explain why part (a) of this exercise agrees with the case $p = -2$ of the power formula.

6. Show that, if $h(x) = \sin(x)$, then $h'(x) = \cos(x)$. Hint: using the trigonometric identity

$$\sin(a + b) = \sin(a)\cos(b) + \cos(a)\sin(b),$$

we find that

$$h'(x) = \lim_{\Delta x \to 0} \frac{\sin(x + \Delta x) - \sin(x)}{\Delta x} = \lim_{\Delta x \to 0} \frac{\sin(x)\cos(\Delta x) + \cos(x)\sin(\Delta x) - \sin(x)}{\Delta x}.$$

If you collect some terms in the numerator, then you can use the limit formulas

$$\lim_{\Delta x \to 0} \frac{\sin(\Delta x)}{\Delta x} = 1 \quad \text{and} \quad \lim_{\Delta x \to 0} \frac{\cos(\Delta x) - 1}{\Delta x} = 0.$$

(See Exercises 14 and 15 in Sect. 1.7.)

7. Show that $\dfrac{d}{dx}[\cos(x)] = -\sin(x)$. The process is similar to that of Exercise 6 above, except that, in the present case, you'll want to use the trigonometric identity

$$\cos(a+b) = \cos(a)\cos(b) - \sin(a)\sin(b).$$

8. Show that $\dfrac{d}{dx}[\tan(x)] = \sec^2(x)$. Hint: use the fact, which may be shown using the trigonometric identities given in the preceding two exercises, that

$$\tan(x + \Delta x) - \tan(x) = \frac{\sin(\Delta x)\sec(x)}{\cos(\Delta x)\cos(x) - \sin(\Delta x)\sin(x)}.$$

You may also want to use one of the limit formulas from Exercise 6 above, and the facts that $\sin(0) = 0$ and $\cos(0) = 1$.

2.3.3.2 Part 2: Differentiation Using Rules and Formulas

9. Find the indicated derivatives, using rules and formulas from this section.

(a) $f'(x)$ if $f(x) = 3x^7 - 0.3x^4 + \pi x^3 - 17$

(b) $\dfrac{d}{dx}\left[\sqrt{3}\sqrt{x} + \dfrac{7}{x^5}\right]$

(c) $h'(w)$ if $h(w) = \dfrac{w^8}{12} - \sin(w) + \dfrac{1}{3w^2}$

(d) $\dfrac{d}{du}\left[\dfrac{4\cos(u)}{5} - \dfrac{3\tan(u)}{8} + \sqrt[3]{u}\right]$

(e) $V'(s)$ if $V(s) = \sqrt[4]{16} - \sqrt[4]{s}$

(f) The derivative with respect to z of $F(z) = \sqrt{7} \cdot 2^z + (1/2)^z$

(g) $P'(t)$ if $P(t) = -\dfrac{a}{2}t^2 + v_0 t + d_0$ (a, v_0, and d_0 are constants)

10. Use Sage (or another computer graphing utility) for this exercise.

For *each* of the parts (a), (b), (c), (d) of this exercise, graph both of the following functions on the same set of axes:

- The function $g(x) = (f(x + .01) - f(x))/0.01$ that estimates the slope of the graph of f at x;
- The function $h(x) = f'(x)$, where you use the differentiation rules to find f'.

(a) $f(x) = x^4$ on $-1 \le x \le 1$.

(b) $f(x) = x^{-1}$ on $1 \le x \le 8$.

(c) $f(x) = \sqrt{x}$ on $.25 \le x \le 9$.

(d) $f(x) = \sin(x)$ on $0 \le x \le 2\pi$.

The graphs of g and h should look similar in each case. Explain why this should be so.

11. In each case below, find a function $f(x)$ whose derivative $f'(x)$ is:

(a) $f'(x) = 12x^{11}$.
(b) $f'(x) = 5x^7$.
(c) $f'(x) = \cos(x) + \sin(x)$.
(d) $f'(x) = ax^2 + bx + c$.
(e) $f'(x) = 0$.
(f) $f'(x) = \dfrac{5}{\sqrt{x}}$.

12. What is the slope of the graph of $y = x - \sqrt{x}$ at $x = 4$? At $x = 100$? At $x = 10000$?
13. Let $f(x) = x^3 - 3x$. Write down a formula for $f'(x)$. Then answer the following, without using any graphing software.

(a) For which values of x is $f(x)$ increasing? Hint: $3x^2 - 3 = 3(x - 1)(x + 1)$. The right hand side will be positive when $x - 1$ and $x + 1$ are either both positive or both negative.
(b) For which values of x is $f(x)$ decreasing?
(c) At what points is the graph of $f(x)$ horizontal?
(d) At what point is the graph of $f(x)$ falling most steeply? Hint: at this point, the slope of the graph is as negative as it can be.
(e) Make a rough sketch, by hand, of a graph of $f(x) = x^3 - 3x$ that reflects all of these results.

14.

(a) Use Sage or a similar utility to sketch the graph of the function $y = 2x + \dfrac{2}{x}$ on the interval $0.2 \le x \le 3$.
(b) Where is the lowest point on that graph? Give the value of the x-coordinate *exactly*. Hint: at this low point, the tangent line is horizontal.

15. What is the slope of the graph of $y = \sin(x) + \cos(x)$ at $x = \pi/4$?
16. Write down two quadratic polynomials $f(x)$ and $g(x)$ that have the same derivative. (A quadratic polynomial is a polynomial of degree two. See Example 2.3.3(b) above.) Supply a computer graph of these two functions, both graphed on the same set of axes.
17. A ball is held motionless and then dropped from the top of a 200 foot tall building. After t seconds have passed, the distance from the ground to the ball is $d = f(t) = -16t^2 + 200\,\text{ft}$.

(a) Find a formula for the velocity $v = f'(t)$ of the ball after t seconds. Check that your formula agrees with the given information that the initial velocity of the ball is $0\,\text{ft/s}$.
(b) Use a computer to draw graphs of both the velocity and the distance as functions of time. What time interval makes physical sense in this situation? (For example,

does $t < 0$ make sense? Does the distance formula make sense after the ball hits the ground?)

(c) At what time does the ball hit the ground? What is its velocity then?

18. A second ball is tossed straight up from the top of the same building with a velocity of 10 feet per second. After t seconds have passed, the distance from the ground to the ball is $d = f(t) = -16t^2 + 10t + 200$ ft.

(a) Find a formula for the velocity of the second ball. Does the formula agree with given information that the initial velocity is $+10$ feet per second? Compare the velocity formulas for the two balls; how are they similar, and how are they different?

(b) Use a computer to draw graphs of both the velocity and the distance as functions of time. What time interval makes physical sense in this situation?

(c) Use your graph to answer the following questions:

(i) During what period of time is the ball rising?

(ii) During what period of time is it falling?

(iii) When does it reach the highest point of its flight?

(iv) How high does the ball rise?

19. A steel ball is rolling along a 20-inch long straight track so that its distance from the midpoint of the track (which is 10 in. from either end) is $d = 3 \sin t$ inches after t seconds have passed. (Think of the track as aligned from left to right. Positive distances mean the ball is to the right of the center; negative distances mean it is to the left.)

(a) Find a formula for the velocity of the ball after t seconds. What is happening when the velocity is positive; when it is negative; when it equals zero? Write a sentence or two describing the motion of the ball.

(b) How far from the midpoint of the track does the ball get? How can you tell?

(c) How fast is the ball going when it is at the midpoint of the track? Does it ever go faster than this? How can you tell?

2.4   The Chain Rule

2.4.1   Leibniz Notation for Derivatives

In this section, we wish to understand how to differentiate chains of functions, like $\sin(x^2)$ (which equals $f(g(x))$, with $f(x) = \sin(x)$ and $g(x) = x^2$), $\sqrt{3 + \tan(t)}$ (which is the chain of $f(t) = \sqrt{t}$ with $g(t) = 3 + \tan(t)$), and so on. To do so, it will be useful to first introduce yet another way of writing derivatives.

Definition 2.4.1 (*Leibniz notation for the derivative*) If $y = f(x)$ is a differentiable function, then we write

$$\frac{dy}{dx}$$

for the derivative $f'(x)$.

For example:

$$\text{If} \quad y = x^9 + x^{3/8} - \frac{\tan(x)}{5}, \quad \text{then} \quad \frac{dy}{dx} = 9x^8 + \frac{3}{8}x^{-5/8} - \frac{\sec^2(x)}{5};$$

$$\text{If} \quad r = 5^q + q^5, \quad \text{then} \quad \frac{dr}{dq} = \ln(5)5^q + 5q^4,$$

and so on.

So if $y = f(x)$ is differentiable, then $f'(x)$, $\frac{d}{dx}[f(x)]$, and $\frac{dy}{dx}$ all mean the same thing. A couple of observations are worth making:

1. $\frac{dy}{dx}$ **is not** a fraction; it's a derivative. BUT:

2. $\frac{dy}{dx}$ **is** *a limit of fractions*, since, by definition of the derivative,

$$\frac{dy}{dx} = \lim_{\Delta x \to 0} \frac{\Delta y}{\Delta x}. \tag{2.16}$$

(Here as usual, if $y = f(x)$, then Δy denotes $f(x + \Delta x) - f(x)$.)

One might think of formula (2.16), informally, as saying "As $\Delta x \to 0$, the Δ's become d's." This is informal because it's only true as far as the above *notation* goes. It's not true about the actual quantities Δx, Δy, dx, and dy. The latter two aren't even quantities *per se*; they're just pieces of the symbol dy/dx. Still, this way of thinking is suggestive. In particular, as we will see in this section, it's quite useful in understanding derivatives of chains.

2.4.2   The Chain Rule, First Version

In Sect. 2.3, we saw that the derivative of a sum is the sum of the derivatives (the sum rule), and that the derivative of a constant multiple is the corresponding constant multiple of the derivative (the constant multiple rule). Now what about chains, or compositions? Is the derivative of a chain equal to the chain of the derivatives? As the following example shows, the answer is *no*, though there *is* a nice "algebra" to differentiation of chains.

Example 2.4.1   At a particular point in time on the elliptical trainer, the monitor says that 0.31 calories (cal) per step are being burned, and steps are being climbed at a rate of 40 per minute (min). What is the time rate of energy expenditure at this point in time, in calories per minute?

Solution. We can answer by "following the units." That is: calories per minute equals calories per step times steps per minute. So the rate of energy expenditure, in calories per minute, equals

$$0.31 \, \frac{\text{cal}}{\text{step}} \cdot 40 \, \frac{\text{step}}{\text{min}} = 0.31 \cdot 40 \, \frac{\text{cal}}{\text{min}} = 12.4 \, \frac{\text{cal}}{\text{min}}.$$

Note that, in the above example, we *are* considering a chain of functions: the number y of calories burned is a function of the number u of steps taken, and u is a function of time x, so y itself is, ultimately, also a function of time x.

Moreover, the above example illustrates the fact that *the rate of change of y with respect to x equals the rate of change of y with respect to u **times** the rate of change of u with respect to x*. This can be summarized quite nicely in Leibniz notation:

$$\boxed{\frac{dy}{dx} = \frac{dy}{du} \cdot \frac{du}{dx}}$$

The chain rule, first version

Very roughly speaking, the chain rule says that "the derivative of the chain equals the product of the derivatives." However, one needs to be a bit careful about such an interpretation. We'll discuss this further in the next subsection, after introduction of our second version of the chain rule.

Note that we haven't actually proved the chain rule. We hope that the above example makes it believable, but those who still need convincing might consider the following. First of all, we have a simple formula for *average* rates of change:

$$\frac{\Delta y}{\Delta x} = \frac{\Delta y}{\Delta u} \cdot \frac{\Delta u}{\Delta x}. \tag{2.17}$$

(This is true just by algebra: cancelling a Δu top and bottom on the right-hand side gives the left-hand side.) But if we apply, to Eq. (2.17), the philosophy that "Δ's become d's as $\Delta x \to 0$," then we get the above chain rule exactly. (This argument is not completely rigorous. But it could be made so with just a few additional details.)

Here are some further examples.

Example 2.4.2 (a) Find:

(i) $\dfrac{dy}{dx}$ if $y = \cos(u)$ and $u = \sin(x)$

(ii) $\dfrac{dz}{dw}$ if $z = 3^v$ and $v = w^2 - 7w$

(iii) $\dfrac{dy}{dx}$ if $y = \sin(x^2)$

(iv) $\dfrac{d}{dt}\left[\sqrt{3 + \tan(t)}\right]$

(b) A spherical snowball is melting, in such a way that its radius is decreasing at 0.75 centimeters (cm) per minute (min). How fast is the surface area of the snowball melting when the radius is 10 cm?

Solution. (a)(i) We have

$$\frac{dy}{dx} = \frac{dy}{du} \cdot \frac{du}{dx} = \frac{d}{du}[\cos(u)] \cdot \frac{d}{dx}[\sin(x)] = -\sin(u) \cdot \cos(x).$$

Now we are asking for a derivative with respect to x, so our answer should be in terms of x. Since $u = \sin(x)$, the above result gives us

$$\frac{dy}{dx} = -\sin(\sin(x))\cos(x).$$

(ii) $\dfrac{dz}{dw} = \dfrac{dz}{dv} \cdot \dfrac{dv}{dw} = \dfrac{d}{dv}[3^v] \cdot \dfrac{d}{dw}[w^2 - 7w] = \ln(3)3^v \cdot (2w - 7) = \ln(3)(2w - 7)$
$\cdot 3^{w^2 - 7w}$.

(iii) We are not explicitly given a "function u in the middle," so we need to introduce one. That is, we write $y = \sin(u)$ where $u = x^2$. Then

$$\frac{dy}{dx} = \frac{dy}{du} \cdot \frac{du}{dx} = \frac{d}{du}[\sin(u)] \cdot \frac{d}{dx}[x^2] = \cos(u) \cdot 2x = 2x\cos(x^2).$$

(iv) Let $y = \sqrt{3 + \tan(t)}$: we write $y = \sqrt{u}$ where $u = 3 + \tan(t)$. Then

$$\frac{d}{dt}\left[\sqrt{3 + \tan(t)}\right] = \frac{dy}{dt} = \frac{dy}{du} \cdot \frac{du}{dt} = \frac{d}{du}\left[\sqrt{u}\right] \cdot \frac{d}{dt}[3 + \tan(t)]$$

$$= \frac{1}{2\sqrt{u}} \cdot (0 + \sec^2(t)) = \frac{\sec^2(t)}{2\sqrt{3 + \tan(t)}}.$$

(b) The surface area S of a sphere is given in terms of its radius r by the formula $S = 4\pi r^2$. If we were interested in the rate of change of S with respect to r, this would be easy: the power formula and the constant multiple rule would give us $dS/dr = d[4\pi r^2]/dr = 4\pi \cdot 2r = 8\pi r$. But because we want the rate of change of S *with respect to time* t (as is indicated by the phrase "how fast"), there is an extra step.

Specifically, the chain rule gives

$$\frac{dS}{dt} = \frac{dS}{dr} \cdot \frac{dr}{dt} = 8\pi r \frac{dr}{dt}. \tag{2.18}$$

We are given that $dr/dt = 0.75$ cm/min, so when $r = 10$ cm we have, by Eq. (2.18),

$$\frac{dS}{dt} = 8\pi \cdot 10 \text{ cm} \cdot 0.75 \frac{\text{cm}}{\text{min}} = 188.496 \frac{\text{cm}^2}{\text{min}}.$$

Note that the units work out: square centimeters per minute are the correct units for a rate of change of area (in centimeters) with respect to time (in minutes).

2.4.3 The Chain Rule, Second Version

The situation embodied by parts (a)(iii) and (a)(iv) of Example 2.4.2 is typical, in that the chains there were given to us in "final" form. That is, the two simpler functions being "chained together" to get the more complex one, whose derivative we wanted, were not specified explicitly. Or to put it another way: we were not explicitly given an "intermediate function" u. Our next version of the chain rule allows us to differentiate chains like this, without having to *explicitly write down* an auxiliary function like u.

To see how this works, let $y = f(g(x))$. We can write $y = f(u)$ where $u = g(x)$. Then by the above chain rule,

$$\frac{dy}{dx} = \frac{dy}{du} \cdot \frac{du}{dx} = f'(u) \cdot g'(x). \tag{2.19}$$

But again, $y = f(g(x))$ and $u = g(x)$; substituting these facts into Eq. (2.19) gives

$$\boxed{\frac{d}{dx}[f(g(x))] = f'(g(x)) \cdot g'(x)}$$

The chain rule, second version

Note that, although we needed a variable u to *arrive at* this version of the chain rule, no such variable appears explicitly in the final result.

This second version of the chain rule says, roughly: to differentiate a chain, follow these two steps.

Step 1. Differentiate the outer function, and into this derivative, substitute the inner function.
Step 2. Multiply the result of Step 1 by the derivative of the inner function.

Steps for using the chain rule, second version

Example 2.4.3 Find:

(a) $h'(x)$ if $h(x) = (x^2 + 45x)^{37}$.

(b) $\dfrac{d}{dz}[\cos(3z)]$.

Solution. (a) {Implicitly, we are thinking that our "outer" function is the 37th power function (that is, our outer function is $f(u) = u^{37}$), and our "inner" function is $g(x) = x^2 + 45x$. The derivative of the outer function is 37 times the 36th power function (that is, $f'(u) = 37u^{36}$), while the derivative of the inner function is given by $g'(x) = 2x + 45$. So by the above chain rule—or, equivalently, by the two-step process described below it—we have the following.}

$\qquad$ $h'(x) = \{$ the derivative of the other function, evaluated at the inner function,

$\qquad\qquad$ · the derivative of the inner function $\}$

$\qquad\qquad = 37(x^2 + 45)^{36}(2x + 45).$

$\quad$ (b) {Our outer function is the cosine function, whose derivative is minus the sine function. Our inner function is $3z$. So the chain rule gives us the following.}

$$\frac{d}{dz}[\cos(3z)] = -\sin(3z)\frac{d}{dz}[3z] = -3\sin(3z).$$

In the above example, everything in curly braces (both in the preambles to the computations and in the computations themselves) is meant implicitly; it indicates the thought processes involved, but would generally not need to be written out *per se*. This is the point of our second version of the chain rule: it allows us to do a certain amount of "bookkeeping"—identifying inner and outer functions—in our head, rather than on paper.

The following examples further illustrate the idea. (You should do the mental bookkeeping of identifying outer and inner functions.)

Example 2.4.4 Find:

(a) $\dfrac{d}{dz}[\sin(\sin(z))]$.

(b) $H'(1)$ if $H(t) = Q(2t^2 + 3t)$ and $Q'(5) = 4$.

(c) $\dfrac{d}{dz}[\sin(\sin(\sin(z)))]$.

Solution. (a) $\dfrac{d}{dz}[\sin(\sin(z))] = \cos(\sin(z))\dfrac{d}{dz}[\sin(z)] = \cos(\sin(z))\cos(z).$

$\quad$ (b) We differentiate first, and *then* evaluate at the particular point $t = 1$. By the chain rule, we have

$$H'(t) = Q'(2t^2 + 3t)\frac{d}{dt}[2t^2 + 3t] = Q'(2t^2 + 3t) \cdot (4t + 3).$$

Substituting $t = 1$ then gives

$$H'(1) = Q'(2 \cdot 1^2 + 3 \cdot 1) \cdot (4 \cdot 1 + 3) = Q'(5) \cdot 7 = 4 \cdot 7 = 28.$$

(c) Applying the chain rule twice in succession gives

$$\frac{d}{dz}[\sin(\sin(\sin(z)))] = \cos(\sin(\sin(z)))\frac{d}{dz}[\sin(\sin(z))] = \cos(\sin(\sin(z)))\cos(\sin(z))\frac{d}{dz}[\sin(z)]$$

$$= \cos(\sin(\sin(z)))\cos(\sin(z))\cos(z).$$

(Here, we first thought of our outer function as the sine function, and our inner function as $\sin(\sin(z))$. Of course, Step 2 of the chain rule tells us that we need to differentiate this inner function; to do that, we applied the chain rule *again*, since the inner function is itself a chain.)

As part (c) of the above example illustrates, our second version of the chain rule is particularly convenient when we have chains of chains (or chains of chains of chains, and so on). Had we attempted this same exercise using our first version of the chain rule, we would have needed to introduce two auxiliary variables: that is, we would have had to write, say, $y = \sin(u)$ where $u = \sin(v)$ and $v = \sin(x)$. The second version of the chain rule allows us to do this kind of "unchaining" behind the scenes.

Warning. We noted in the previous subsection that, in a sense, the chain rule expresses the derivative of a chain as a product of derivatives. This is true, but only if properly interpreted. That is: the chain rule does *not* say that the derivative of $f(g(x))$ is $f'(x)$ times $g'(x)$. It *does* say that the derivative of $f(g(x))$ is $f'(g(x))$ times $g'(x)$. So: "the derivative of a chain equals the product of the derivatives, as long as the latter two are evaluated at the appropriate, distinct places."

2.4.4 Exercises

2.4.4.1 Part 1: The Chain Rule, First Version

For these exercises, you should refer back to the subsection "The chain rule, first version" above, and especially Example 2.4.2.

1. Use the chain rule to find dy/dx, when y is given as a function of x in the following way.

(a) $y = 5u - 3$, where $u = 4 - 7x$.
(b) $y = \sin u$, where $u = 4 - 7x$.
(c) $y = \tan u$, where $u = x^3$.
(d) $y = 10^u$, where $u = x^2$.
(e) $y = u^4$, where $u = x^3 + 5$.

2. A cube of sidelength ℓ has surface area $S = 6\ell^2$.
   Suppose that, at the moment when the sides of a cube are 5 in. in length, these sides are growing at a rate of 3 inches per hour. How fast is the surface area increasing, in square inches per hour?

3. An explorer is marooned on an iceberg. The top of the iceberg is shaped like a square, with sides of length 100 ft. The length of the sides is shrinking at the rate of two feet per day. How fast is the area of the top of the iceberg shrinking? Assuming the sides continue to shrink at the rate of two feet per day, what will be the dimensions of the top of the iceberg in five days? How fast will the area of the top of the iceberg be shrinking then?

4. Suppose the iceberg of Exercise 3 is shaped like a cube. How fast is the volume of the cube shrinking when the sides have length 100 ft? How fast after five days? (A cube of sidelength ℓ has volume $V = \ell^3$.)

5. (Note: only parts (c) and (d) of this exercise involve the chain rule.) If the radius of a spherical balloon is r inches, its volume V is $\frac{4}{3}\pi r^3$ cubic inches.

   (a) Find the rate of change dV/dr of V with respect to r.
   (b) At what rate does the volume increase, in cubic inches per inch, when the radius is 4 in.?
   (c) Use the chain rule to express the rate of change of volume V, *with respect to time* t, in terms of the radius r, and the rate of change dr/dt of the radius with respect to time.
   (d) Suppose someone is inflating the balloon at the rate of 10 cubic inches of air per second. If the radius is 4 in., at what rate is it increasing, in inches per second?

2.4.4.2 Part 2: The Chain Rule, Second Version

For these exercises, you should refer back to the subsection "The chain rule, second version" above, and especially Examples 2.4.3 and 2.4.4.

6. Find the derivatives of the following functions.

   (a) $F(x) = (9x + 6x^3)^5$.
   (b) $G(w) = \sqrt{4w^2 + 1}$.
   (c) $R(x) = \dfrac{1}{1-x}$. Hint: $\dfrac{1}{1-x} = (1-x)^{-1}$.

(d) $D(z) = 3 \tan\left(\dfrac{1}{z}\right).$

(e) $optimus(t) = \cos(2^t).$

(f) $whatever(x) = 5^{1/x}.$

(g) $twin(x) = \cos(-27x).$

7. Apply the chain rule more than once to find each of the following derivatives. (See Example 2.4.4(c).)

(a) $dog(w) = \sin^2(w^3 + 1)$. Hint: first write $\sin^2(w^3 + 1) = (\sin(w^3 + 1))^2$.

(b) $q(y) = \tan(\sin(\cos(y))).$

(c) $R(x) = 3x + (x^2 + (7x^3 + 5)^2)^3.$

8. Let $S(w) = \sqrt{(4w^2 + 1)^3}$. Find $S'(w)$ in two ways:

   (i) Write $\sqrt{(4w^2 + 1)^3} = (4w^2 + 1)^{3/2}$ and use the chain rule once.

   (ii) Write $\sqrt{(4w^2 + 1)^3} = ((4w^2 + 1)^3)^{1/2}$ and use the chain rule twice.

   Make sure to show that your two answers are the same.

9. Let $f(t) = t^2 + 2t$ and $g(t) = 5t^3 - 3$. Determine all of the following: $f'(t)$, $g'(t)$, $g(f(t))$, $f(g(t))$, $g'(f(t))$, $f'(g(t))$, $\dfrac{d}{dt}[f(g(t))]$, $\dfrac{d}{dt}[g(f(t))]$.

2.4.4.3 Part 3: Particular Values

For the following exercises remember that, to evaluate a derivative at a point, you differentiate first, and *then* plug in the point in question. See Example 2.4.4(b) above.

10. If $h(x) = (f(x))^6$ where f is some function satisfying $f(93) = 2$ and $f'(93) = -4$, what is $h'(93)$?

11. If $H(x) = F(x^2 - 4x + 2)$ where F is some function satisfying $F'(2) = 3$, what is $H'(4)$?

12. If $f(x) = (1 + x^2)^5$, what are the numerical values of $f'(0)$ and $f'(1)$?

13. If $h(t) = \sin(\sin t)$, what are the numerical values of $h'(0)$ and $h'(\pi)$?

14. If $k(t) = \sin(\pi \sin \pi t)$, what are the numerical values of $k'(0)$ and $k'(1)$?

2.4.4.4 Part 4: Miscellaneous

15. (a) What is the derivative of $f(x) = 2^{-x^2}$?

    (b) Sketch the graphs of f and its derivative on the interval $-2 \le x \le 2$.

    (c) For what values(s) of x is $f'(x) = 0$? What is true about the graph of f at the corresponding points?

   (d) Where does the graph of f have positive slope, and where does it have negative slope?

16. (a) With a graphing utility, find the point x where the function $y = 1/(3x^2 - 5x + 7)$ takes its maximum value. Obtain the numerical value of x accurately to two decimal places.

   (b) Find the derivative of $y = 1/(3x^2 - 5x + 7)$, and determine where it takes the value 0.

   (c) Using part (b), find the *exact* value of x where $y = 1/(3x^2 - 5x + 7)$ takes its maximum value. Hint: for a quotient a/b to be zero, the numerator must be zero.

17. (a) Find a function $f(x)$ for which $f'(x) = 3x^2(5 + x^3)^{10}$. A useful way to proceed is to guess. For instance, you might guess $f(x) = (5 + x^3)^{11}$. Differentiate $f(x)$ and see what you get; see if you can use this information to modify $f(x)$, to get the correct answer.

   (b) Find a function $p(x)$ for which $p'(x) = x^2(5 + x^3)^{10}$.

18. Find a function $g(t)$ for which $g'(t) = t/\sqrt{1 + t^2}$. Hint: guess $g(t) = \sqrt{1 + t^2}$. As in Exercise 17 above, differentiate $g(t)$, and then modify your guess to correct it.

19. Find a function $h(x)$ for which $h'(x) = x7^{x^2}$. Hint: guess $h(x) = 7^{x^2}$, then proceed as in the previous two exercises.

2.5   More Differentiation Rules

A mathematical process is said to *commute* with differentiation if it doesn't matter whether you differentiate before, or after, you perform that process.

We've seen, for example, that differentiation commutes with addition of functions, and with multiplication of functions by constants. That is: the sum rule tells us that the derivative of the sum is the sum of the derivatives; the constant multiple rule tells us that, if you multiply a function by a *constant* and then differentiate the result, then you get the same thing as if you differentiate first, and then multiply. (Food for thought: which of the latter two sequences of processes corresponds to which side of the constant multiple rule?) On the other hand, differentiation *does not* commute with composition of functions: the derivative of a chain is not, generally, equal to the chain of the derivatives.

In this section, we explore some additional processes that, like the chain rule, *do not* commute with differentiation.

2.5.1   The Product Rule

To see right away that the derivative of the product of two functions does not, in general, equal the product of the two derivatives of those functions, let $f(x) = x^2$ and $g(x) = x^3$. Then the derivative of the product is

$$\frac{d}{dx}[f(x)g(x)] = \frac{d}{dx}[x^2 \cdot x^3] = \frac{d}{dx}[x^5] = 5x^4,$$

while the product of the derivatives equals

$$\frac{d}{dx}[f(x)] \cdot \frac{d}{dx}[g(x)] = \frac{d}{dx}[x^2] \cdot \frac{d}{dx}[x^3] = 2x \cdot 3x^2 = 6x^3.$$

And certainly $5x^4 \neq 6x^3$ (except in the special cases $x = 0$ and $x = 6/5$)!

To understand how products *do* behave, in terms of change in their factors, consider a product $r = pq$ of two quantities p and q, both dependent on a variable x. How does r respond to changes in both p and q?

To answer, suppose p changes by an amount Δp, and q by an amount Δq. Then the corresponding change Δr in $r = pq$ is given by

$$\begin{aligned}
\Delta r &= \text{new } r - \text{old } r \\
&= (p + \Delta p)(q + \Delta q) - pq \\
&= pq + p \cdot \Delta q + \Delta p \cdot q + \Delta p \cdot \Delta q - pq \\
&= p \cdot \Delta q + \Delta p \cdot q + \Delta p \cdot \Delta q.
\end{aligned} \tag{2.20}$$

This calculation is modeled, and encapsulated, in Fig. 2.6. Here, the change in the area of the rectangle, as its sides grow from lengths p and q to lengths $p + \Delta p$ and $q + \Delta q$, equals the sum of the areas of the three "extra" bits.

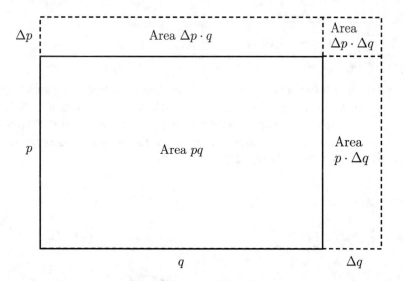

Fig. 2.6 The change in a product, modeled geometrically

Of course, Eq. (2.20) holds in any setting entailing a product of two quantities p and q, not just a geometric setting. But the geometric model is a suggestive one.

Now with this general setting in mind, let's divide both sides of Eq. (2.20) through by Δx; we get

$$\frac{\Delta r}{\Delta x} = p\frac{\Delta q}{\Delta x} + q\frac{\Delta p}{\Delta x} + \Delta p\frac{\Delta q}{\Delta x}. \tag{2.21}$$

We let Δx approach zero. The average rates of change become instantaneous rates of change, and $\Delta p \to 0$, so Eq. (2.21) yields

$$\frac{dr}{dx} = p\frac{dq}{dx} + q\frac{dp}{dx} + 0 \cdot \frac{dq}{dx} = p\frac{dq}{dx} + q\frac{dp}{dx},$$

or, writing $p = f(x)$ and $q = g(x)$,

$$\boxed{\frac{d}{dx}[f(x)g(x)] = f(x)g'(x) + g(x)f'(x)}$$

The product rule

In words: the derivative of a product equals the first factor times the derivative of the second, plus the second factor times the derivative of the first.

Example 2.5.1

(a) Differentiate $y = x^2 \cos(x)$.
(b) Find $\dfrac{d}{dx}\left[\sin(x^5 + 4)\cos(2^x)\right]$.
(c) Find $\dfrac{d}{dz}\left[\tan(z^5 3^z)\right]$.
(d) Suppose the per capita yearly energy consumption in a country is currently 800,000 BTU, and, due to energy conservation efforts, it is falling at the rate of 1,000 BTU per year. Suppose too that the population of the country is currently 200,000,000 people, and is rising at the rate of 1,000,000 people per year. Is the total yearly energy consumption of this country rising or falling? By how much?

Solution. (a) We have

$$\frac{dy}{dx} = x^2 \frac{d}{dx}[\cos(x)] + \cos(x)\frac{d}{dx}[x^2] = x^2(-\sin(x)) + \cos(x) \cdot 2x = -x^2 \sin(x) + 2x\cos(x).$$

(b) By the product rule, followed by two applications of the chain rule,

$$\frac{d}{dx}\Big[\sin(x^5+4)\cos(2^x)\Big] = \sin(x^5+4)\frac{d}{dx}\Big[\cos(2^x)\Big] + \cos(2^x)\frac{d}{dx}\Big[\sin(x^5+4)\Big]$$

$$= \sin(x^5+4)(-\sin(2^x))\frac{d}{dx}[2^x] + \cos(2^x)\cos(x^5+4)\frac{d}{dx}[x^5+4]$$

$$= \sin(x^5+4)(-\sin(2^x))\cdot\ln(2)2^x + \cos(2^x)\cos(x^5+4)\cdot 5x^4$$

$$= -\ln(2)2^x\sin(2^x)\sin(x^5+4) + 5x^4\cos(2^x)\cos(x^5+4).$$

(c) By the chain rule, followed by the product rule,

$$\frac{d}{dz}\Big[\tan(z^53^z)\Big] = \sec^2(z^53^z)\frac{d}{dz}\Big[z^53^z\Big]$$

$$= \sec^2(z^53^z)\Big(z^5\frac{d}{dz}[3^z] + 3^z\frac{d}{dz}[z^5]\Big)$$

$$= \sec^2(z^53^z)\Big(z^5\cdot\ln(3)3^z + 3^z\cdot 5z^4\Big)$$

$$= \sec^2(z^53^z)3^z\Big(\ln(3)z^5 + 5z^4\Big).$$

(d) We can model this situation with three functions $C(t)$, $P(t)$, and $E(t)$:

$C(t)$: per capita yearly energy consumption at time t

$P(t)$: population at time t

$E(t)$: total yearly energy consumption at time t

We are interested in whether, and how, $E(t)$ is increasing or decreasing; that is, we are interested in $E'(t)$.

Now total yearly energy consumption equals per capita yearly energy consumption times the number of people in the population; that is,

$$E(t) = C(t)P(t).$$

We differentiate using the product rule, to find that

$$E'(t) = C(t)P'(t) + P(t)C'(t). \tag{2.22}$$

Next: if $t = 0$ represents today, then we are given the two rates of change

$$C'(0) = -1,000 = -10^3 \text{ BTU per person per year, and}$$

$$P'(0) = 1,000,000 = 10^6 \text{ persons per year.}$$

Similarly, we're given $C(0)$ and $P(0)$. So by Eq. (2.22),

$$E'(0) = C(0) P'(0) + P(0)C'(0)$$
$$= (8 \cdot 10^5) \cdot (10^6) + (2 \cdot 10^8) \cdot (-10^3)$$
$$= (8 \cdot 10^{11}) - (2 \cdot 10^{11})$$
$$= 6 \cdot 10^{11} \text{ BTU per year.} \tag{2.23}$$

So the total yearly energy consumption is currently rising at the rate of $6 \cdot 10^{11}$ BTU per year. Note what this means: the growth in the population more than offsets the efforts to conserve energy.

In (2.23), we omitted the units until the very end. But the units do check out: $C(t)$ represents *per capita* yearly energy consumption, so its units are BTU/person. Therefore, the units for $C(0) \cdot P'(0)$ are

$$\frac{\text{BTU}}{\text{person}} \cdot \frac{\text{persons}}{\text{year}} = \frac{\text{BTU}}{\text{year}},$$

and, similarly, the units for $P(0)C'(0)$ are

$$\text{persons} \cdot \frac{\text{BTU/person}}{\text{year}} = \frac{\text{BTU}}{\text{year}}.$$

2.5.2   The Quotient Rule

As a prelude to differentiating general quotients $f(x)/g(x)$ of functions, we first consider the particularly simple case when $f(x) = 1$: that is, we consider derivatives of *reciprocals* $1/g(x)$.

Recall that the derivative of $1/x$ is -1 over the square of x. So, by the chain rule, the derivative of one over a *function* of x is minus one over the square of that function of x, **times** the derivative of that function of x. (See, for example, "Steps for using the chain rule, second version," on Sect. 2.4.3.)

In symbols,

$$\frac{d}{dx}\left[\frac{1}{g(x)} \right] = -\frac{1}{(g(x))^2} \cdot \frac{d}{dx}[g(x)]$$

or, more simply,

$$\boxed{\frac{d}{dx}\left[\frac{1}{g(x)} \right] = -\frac{g'(x)}{(g(x))^2}}$$

The reciprocal rule

Example 2.5.2

(a) Find: (i) $\dfrac{d}{dt}\left[\dfrac{1}{t^2+1}\right]$; (ii) $\dfrac{d}{dt}\left[\dfrac{1}{(t^2+1)(t^2+2)}\right]$.

(b) Use the reciprocal rule (and what you know about derivatives of cosines) to show that
$$\frac{d}{dx}[\sec(x)] = \sec(x)\tan(x).$$

Solution. (a)(i) By the reciprocal rule,

$$\frac{d}{dt}\left[\frac{1}{t^2+1}\right] = -\frac{d[t^2+1]/dt}{(t^2+1)^2} = -\frac{2t}{(t^2+1)^2}.$$

(ii) By the reciprocal rule and the product rule,

$$\frac{d}{dt}\left[\frac{1}{(t^2+1)(t^2+2)}\right] = -\frac{\dfrac{d}{dt}[(t^2+1)(t^2+2)]}{(t^2+1)^2(t^2+2)^2} = -\frac{(t^2+1)\dfrac{d}{dt}[t^2+2] + (t^2+2)\dfrac{d}{dt}[t^2+1]}{(t^2+1)^2(t^2+2)^2}$$

$$= -\frac{(t^2+1)\cdot 2t + (t^2+2)\cdot 2t}{(t^2+1)^2(t^2+2)^2} = -\frac{2t(t^2+1+t^2+2)}{(t^2+1)^2(t^2+2)^2} = -\frac{2t(2t^2+3)}{(t^2+1)^2(t^2+2)^2}.$$

(b) By the definitions of $\sec(x)$ and $\tan(x)$, by the reciprocal rule, and by some algebra, we have

$$\frac{d}{dx}[\sec(x)] = \frac{d}{dx}\left[\frac{1}{\cos(x)}\right] = -\frac{-\sin(x)}{\cos^2(x)} = \frac{\sin(x)}{\cos^2(x)} = \frac{1}{\cos(x)}\cdot\frac{\sin(x)}{\cos(x)} = \sec(x)\tan(x).$$

In our derivation of the reciprocal rule above, we began by noting that $\dfrac{d}{dx}\left[\dfrac{1}{x}\right] = -\dfrac{1}{x^2}$. But this latter formula is just the case $p = -1$ of the power formula. So the reciprocal rule is itself a special case of the power formula, combined with the chain rule.

We now turn to differentiation of general quotients. The key is to note that a quotient *is* a kind of product: in general, $a/b = (1/b)\cdot a$. So, by the above product and reciprocal rules,

$$\frac{d}{dx}\left[\frac{f(x)}{g(x)}\right] = \frac{d}{dx}\left[\frac{1}{g(x)}\cdot f(x)\right] = \frac{1}{g(x)}\frac{d}{dx}[f(x)] + f(x)\frac{d}{dx}\left[\frac{1}{g(x)}\right]$$

$$= \frac{1}{g(x)}\cdot f'(x) + f(x)\cdot\left(-\frac{g'(x)}{(g(x))^2}\right) = \frac{f'(x)}{g(x)} - \frac{f(x)g'(x)}{(g(x))^2}. \tag{2.24}$$

To clean this up, we multiply top and bottom of the first quotient on the right-hand side of Eq. (2.24) by $g(x)$, to get a common denominator. We get

$$\frac{d}{dx}\left[\frac{f(x)}{g(x)}\right] = \frac{g(x)f'(x)}{(g(x))^2} - \frac{f(x)g'(x)}{(g(x))^2}$$

or, finally,

$$\boxed{\frac{d}{dx}\left[\frac{f(x)}{g(x)}\right] = \frac{g(x)f'(x) - f(x)g'(x)}{(g(x))^2}}$$

The quotient rule

Example 2.5.3

(a) Find: (i) $\dfrac{d}{dx}\left[\dfrac{2^x}{2^x+1}\right]$; (ii) $\dfrac{d}{dx}\left[\dfrac{x\cos(x)}{\sin(x^2)}\right]$.

(b) Use the reciprocal rule (and what you know about derivatives of cosines and sines) to
show that $\dfrac{d}{dx}[\cot(x)] = -\csc^2(x)$.

Solution. (a)(i) By the quotient rule,

$$\frac{d}{dx}\left[\frac{2^x}{2^x+1}\right] = \frac{(2^x+1)\dfrac{d}{dx}[2^x] - 2^x\dfrac{d}{dx}[2^x+1]}{(2^x+1)^2} = \frac{(2^x+1)\cdot\ln(2)2^x - 2^x\cdot\ln(2)2^x}{(2^x+1)^2}$$

$$= \frac{\ln(2)2^x(2^x+1-2^x)}{(2^x+1)^2} = \frac{\ln(2)2^x}{(2^x+1)^2}.$$

(ii) By the quotient rule, the product rule, and the chain rule,

$$\frac{d}{dx}\left[\frac{x\cos(x)}{\sin(x^2)}\right] = \frac{\sin(x^2)\dfrac{d}{dx}[x\cos(x)] - x\cos(x)\dfrac{d}{dx}[\sin(x^2)]}{\sin^2(x^2)}$$

$$= \frac{\sin(x^2)\left(x\dfrac{d}{dx}[\cos(x)] + \cos(x)\dfrac{d}{dx}[x]\right) - x\cos(x)\cos(x^2)\dfrac{d}{dx}[x^2]}{\sin^2(x^2)}$$

$$= \frac{\sin(x^2)\left(-x\sin(x) + \cos(x)\right) - 2x^2\cos(x)\cos(x^2)}{\sin^2(x^2)}$$

$$= \frac{-x\sin(x)\sin(x^2) + \cos(x)\sin(x^2) - 2x^2\cos(x)\cos(x^2)}{\sin^2(x^2)}.$$

(b) By the definitions of $\cot(x)$ and $\csc(x)$, the quotient rule, the trigonometric identity

$$\cos^2(x) + \sin^2(x) = 1,$$

and some algebra, we have

$$\frac{d}{dx}[\cot(x)] = \frac{d}{dx}\left[\frac{\cos(x)}{\sin(x)}\right] = \frac{\sin(x)\frac{d}{dx}[\cos(x)] - \cos(x)\frac{d}{dx}[\sin(x)]}{\sin^2(x)}$$

$$= \frac{\sin(x)(-\sin(x)) - \cos(x)\cos(x)}{\sin^2(x)}$$

$$= \frac{-(\sin^2(x) + \cos^2(x))}{\sin^2(x)} = \frac{-1}{\sin^2(x)} = -\left(\frac{1}{\sin(x)}\right)^2 = -\csc^2(x).$$

2.5.3   Summary of Differentiation Rules

Here are all the differentiation rules that we have encountered thus far (Table 2.2).

Here c is an arbitrary real number, and $f(x)$ and $g(x)$ are any differentiable functions. (The reciprocal and quotient rules hold only for those x such that $g(x) \neq 0$.)

The reciprocal rule may be considered a special case of the quotient rule (see the exercises below).

Table 2.2 A short table of differentiation rules

Function $y = h(x)$	Derivative $\dfrac{dy}{dx} = h'(x) = \dfrac{d}{dx}[h(x)]$	Name of rule
$cf(x)$	$cf'(x)$	Constant multiple rule
$f(x) + g(x)$	$f'(x) + g'(x)$	Sum rule
$f(g(x))$	$f'(g(x))g'(x)$	Chain rule
$f(x)g(x)$	$f(x)g'(x) + g(x)f'(x)$	Product rule
$\dfrac{1}{g(x)}$	$-\dfrac{g'(x)}{(g(x))^2}$	Reciprocal rule
$\dfrac{f(x)}{g(x)}$	$\dfrac{g(x)f'(x) - f(x)g'(x)}{(g(x))^2}$	Quotient rule

2.5.4   Exercises

2.5.4.1 Part 1: Finding Derivatives

1. Find the derivative of each of the following functions.

(a) $3x^5 - 10x^2 + 8$ (g) $x^2 3^x$ (m) $2\sqrt{x} - \dfrac{1}{\sqrt{x}}$

(b) $(5x^{12} + 2)(\pi - \pi^2 x^4)$ (h) $\cos(x) + 5^x$ (n) $\tan(z)\,(\sin(z) - 5)$

(c) $\sqrt{u} - 3/u^3 + 2u^7$ (i) $\sin(x)/\cos(x)$ (o) $\dfrac{\sin(x)}{x^2}$

(d) $mx + b$ $(m, b$ constant) (j) $5^x \cos(x)$ (p) $64^{\cos(t)}/(5\sqrt[3]{t})$

(e) $0.5\sin(x) + \sqrt[3]{x} + \pi^2$ (k) $\dfrac{2^x}{10 + \sin(x)}$ (q) $4^{x^2 + x 2^x}$

(f) $\dfrac{\pi - \pi^2 x^4}{5x^{12} + 2}$ (l) $\sin(4^x \cos(x))$ (r) $\dfrac{5x^2 + \cos(x)}{7\sqrt{x} + 5}$

2. Suppose f and g are functions and that we are given

$$f(2) = 3, \quad g(2) = 4, \quad g(3) = 2, \quad f'(2) = 2, \quad g'(2) = -1, \quad g'(3) = 17.$$

Evaluate the derivative of each of the following functions at $t = 2$:

(a) $f(t) + g(t)$ (f) $\sqrt{g(t)}$ (d) $f(t)/g(t)$
(b) $5f(t) - 2g(t)$ (g) $t^2 f(t)$ (e) $g(f(t))$
(c) $f(t)g(t)$ (h) $(f(t))^2 + (g(t))^2$ (j) $f(3t - (g(1 + t))^2)$

(k) What additional piece of information would you need to calculate the derivative of $f(g(t))$ at $t = 2$?

3. (a) Use the product rule twice to show that

$$\frac{d}{dx}[f(x)g(x)h(x)] = f(x)g(x)h'(x) + f(x)h(x)g'(x) + g(x)h(x)f'(x).$$

(b) If the length, width, and height of a rectangular box are changing at the rates of 3, 6, and -5 in./min at the moment when all three dimensions happen to be 10 in., at what rate is the volume of the box changing then?

(c) If the length, width, and height of a box are 10 in., 12 in., and 8 in., respectively, and if the length and height of the box are changing at the rates of 3 in./min and -2 in./min, respectively, at what rate must the width be changing to keep the volume of the box constant?

4. Which of the following functions has a derivative which is always positive (except at $x = 0$, where neither the function nor its derivative is defined)? Please explain your answer.

$$1/x \qquad -1/x \qquad 1/x^2 \qquad -1/x^2$$

5. Suppose that the current total yearly energy consumption in a particular country is $16 \cdot 10^{13}$ BTU and is rising at the rate of $6 \cdot 10^{11}$ BTU per year. Suppose that the current population is $2 \cdot 10^8$ people and is rising at the rate of 10^6 people per year. What is the current yearly per capita energy consumption? Is it rising or falling? By how much?

6. The population of a particular country is 15,000,000 people and is growing at the rate of 10,000 people per year. In the same country the per capita yearly expenditure for energy is $1,000 per person and is growing at the rate of $8 per year. What is the country's current total yearly energy expenditure? How fast is the country's total yearly energy expenditure growing?

7. The population of a particular country is 30 million and is rising at the rate of 4,000 people per year. The total yearly personal income in the country is 20 billion dollars, and it is rising at the rate of 500 million dollars per year. What is the current per capita personal income? Is it rising or falling? By how much?

2.5.4.2 Part 2: Deriving Differentiation Rules

8. In this exercise we prove the sum rule: $F(x) = f(x) + g(x)$ implies $F'(x) = f'(x) + g'(x)$.

   (a) Show $F(x + \Delta x) - F(x) = f(x + \Delta x) - f(x) + g(x + \Delta x) - g(x)$.
   (b) Divide by Δx and finish the argument.

9. Prove the reciprocal rule by putting $f(x) = 1$ (that is, $f(x)$ is constant and equal to 1) into the quotient rule.

2.5.4.3 Part 3: Second Derivatives

10. If $y = f(x)$, then the **second derivative** of f is just the derivative of the derivative of f; it is denoted $f''(x)$ or d^2y/dx^2. For example, the second derivative of $f(x) = x^7$ is found by first computing that $f'(x) = 7x^6$, so that $f''(x) = d[7x^6]/dx = 7 \cdot 6x^5 = 42x^5$.

Find the second derivative of each of the following functions.

(a) $f(x) = 4x^3 - 7x^2 - 15x + 11$

(b) $f(t) = 2^{3t-2}$

(c) $f(x) = \sin \omega x$, where ω is a constant

(d) $f(x) = x^2 \cos(x)$

(e) $q(r) = r^3 e^r$

(f) $p(z) = \dfrac{z}{z+1}$

2.5.4.4 Part 4: The Colorado River Problem

Make your answer to this sequence of questions an essay. Identify all the variables you consider (e.g., "A stands for the area of the lake"), and indicate the functional relationships between them ("A depends on time t, measured in weeks from the present"). Identify the derivatives of those functions, as necessary.

The Colorado River—which excavated the Grand Canyon, among others—used to empty into the Gulf of California. It no longer does. Instead, it runs into a marshy area some miles from the Gulf and stops. One of the major reasons for this change is the construction of dams—notably the Hoover Dam. Every dam creates a lake behind it, and every lake increases the total surface area of the river. Since the rate at which water evaporates is proportional to the area of the water surface exposed to air, the lakes along the Colorado have increased the loss of river water through evaporation. Over the years, these losses (in conjunction with other factors, like increased usage by a rapidly growing population) have been significant enough to dry up the river at its mouth.

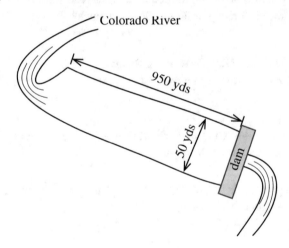

11. Let us analyze the evaporation rate along a river that was recently dammed. Suppose the lake is currently 50 yards wide, and getting wider at a rate of 3 yards per week. As the lake fills, it gets longer, too. Suppose it is currently 950 yards long, and it is extending upstream at a rate of 15 yards per week. Assuming the lake remains approximately rectangular as it grows, find

  (a) the current area of the lake, in square yards;
  (b) the rate at which the surface of the lake is currently growing, in square yards per week.

12. Suppose the lake continues to spread sideways at the rate of 3 yards per week, and it continues to extend upstream at the rate of 15 yards per week.

   (a) Express the area of the lake as a (quadratic!) function of time, where time is measured from the present, in weeks, and where the lake's area is as given in Exercise 20.
   (b) How many weeks will it take for the lake to cover 30 acres (= 145, 200 square yards)?
   (c) At what rate is the lake surface growing when it covers 30 acres?

13. Compare the rates at which the surface of the lake is growing in Exercise 11 (which is the "current" rate) and in Exercise 12(c) (which is the rate when the lake covers 30 acres). Are these rates the same? If they are not, how do you account for the difference? In particular, the width and length grow at fixed rates, so why doesn't the area? Use what you know about derivatives to answer the question.

14. Suppose the local climate causes water to evaporate from the surface of the lake at the rate of 0.22 cubic yards per week, for each square yard of surface. Write a formula that expresses total evaporation per week in terms of area. Use E to denote total evaporation.

15. The lake is fed by the river, and that in turn is fed by rainwater and groundwater from its watershed. (The **watershed**, or basin, of a river is that part of the countryside containing the ponds and streams which drain into the river.) Suppose the watershed provides the lake, on average, with 25,000 cubic yards of new water each week.

    Assuming, as we did in Exercise 11, that the lake widens at the constant rate of 3 yards per week, and lengthens at the rate of 15 yards per week, will the time ever come that the water being added to the lake from its watershed balances the water being removed by evaporation? In other words, will the lake ever stop filling?

2.6   Optimization, Part I: Extreme Points of a Function

As we will see in this and the next section, calculus is a powerful tool for solving **optimization problems**. These are problems whose solutions require finding maximum and/or minimum values of a function.

Such problems occur frequently in many contexts:

- A manufacturer will often seek to maximize profit, which will depend on variables like the cost of the raw materials and the unit price charged per product.
- A soap bubble, stretched across a frame, will form a shape whose surface area is as small as possible, given the constraints imposed by the shape of the frame. This is because the surface area of the soap bubble is proportional to its *surface energy*, and states of minimal energy are stable states. (A search for "minimal soap bubbles" on YouTube will reveal

a variety of interesting and amusing videos.) Many other physical phenomena adhere to similar minimum principles.

- A relationship between variables x and y is sometimes studied by selecting a variety of input values $x_1, x_2, \ldots, x_n$, observing the corresponding outputs $y_1, y_2, \ldots, y_n$, and plotting the resulting points $(x_1, y_1), (x_2, y_2), \ldots, (x_n, y_n)$ in the plane. Such a plot is called a *scatterplot*, and indeed, in reality, the data will often be quite scattered. To make sense of this data, one often tries to model it by a line that approximates the data in some optimal sense. Typically, the line that one uses is the *regression line*, which is the line $y = f(x)$ that that minimizes the sum, over all values of k, of the squares of the differences $y_k - f(x_k)$. That is, the regression line minimizes the sum of the squares of the discrepancies between the actual, observed outputs y_k and the theoretical outputs $f(x_k)$ that would result were the output a linear function of the input. (For this reason, a regression line is sometimes called a "least squares best fit" line.)

- Navigation apps will seek to find a route that minimizes the time it takes to get from a given starting point to a desired destination. Variables here will include distance, speed limits, traffic, road conditions, and so on. Some of these apps will also search for routes that are optimal in other ways—for example, with respect to distance traveled or fuel efficiency.

In many optimization problems, the derivative is the key tool. To see why this is so, we need to look at bit more closely at continuous functions, and their "high" and "low" points.

2.6.1 Extremes and Critical Points

A function defined and continuous on a given interval can have various **local extremes** on that interval. Here, by **local extreme**, we mean a point at which the function "tops out" or "bottoms out" compared to nearby points.

Consider for example, the following continuous function f, whose domain we take (for the moment) to be the interval $(0, \infty)$ of positive real numbers.

This function f, on this domain, has a local extreme at each of the points $x = p, x = q, x = s$, and $x = t$. More specifically, f has a **local minimum** at $x = p$, since $f(p) \le f(x)$ for all points x in the domain of f that are *sufficiently near* the point p. Similarly, f has a local minimum at $x = s$. Further, f has a **local maximum** at $x = q$, since $f(q) \ge f(x)$ for all points x in the domain of f that are *sufficiently near* the point q. Similarly, f has a local maximum at $x = t$.

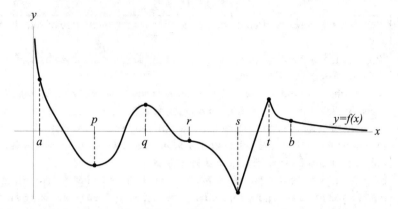

Fig. 2.7 A continuous function f

An important thing to note is that, **at each of these local extremes, the derivative f' is either zero or is undefined.** A point x in the domain of a function f where $f'(x) = 0$ or $f'(x)$ is undefined is called a **critical point** of f.

So we see that, for the above function f, with domain $(0, \infty)$, all local extremes occur at critical points. Note, though, that the converse is not true: the point $x = r$ is a critical point (the graph of f is horizontal at $x = r$, so $f'(r) = 0$), but there is no local minimum or maximum at this point.

We also note that, if the domain of a continuous function contains any endpoints, then local extremes might occur at those points as well. For example: if we take the domain of the above function f to be $[a, b]$, with a and b as in Fig. 2.7, then we see that f has a local maximum at $x = a$ (since $f(a) \geq f(x)$ for all x in the domain of f and sufficiently near the point a), and a local minimum at $x = b$.

In optimization problems, we often want to find the point(s) at which a function attains it **global minimum**, meaning its very smallest value (on its domain), and/or its **global maximum**, meaning its very largest value (on its domain). One way to do so is to first narrow our search by locating all *local* extremes, and then to compare these extremes to each other, to determine which is the largest and which is the smallest.

The above function f, when taken to have domain $(0, \infty)$, has its global minimum at $x = s$, but attains *no* global maximum (since $f(x)$ grows without bounds as x approaches 0 from the right). As a function with domain $[a, b]$, on the other hand, f *does* have a global maximum, at $x = a$ (and again has global minimum at $x = s$).

Figure 2.7 and the ensuing discussion then suggest the following strategy for determining global extremes of continuous functions.

Step 1. Determine the domain of f, and identify the endpoints of this domain, if any.

Step 2. Find any critical points of f, meaning points x in the domain of f where $f'(x) = 0$ or $f'(x)$ is undefined.

Step 3. Suppose the domain of f is an interval of the form $[a, b]$, where a and b are (finite) real numbers. Such a domain is called a **closed, bounded interval**. We may then identify the global extremes of f by simply evaluating f at each of the endpoints and critical points of this domain, and determining at which of these points f attains its smallest and largest values.

Step 4. Suppose the domain of f is *not* a closed, bounded interval. (For example, the domain might be $(0, 1]$, $[0, 1)$, $(0, 1)$, $(-\infty, 1]$, $(0, \infty)$, or $(-\infty, \infty)$.) Then a bit more care is required to determine whether f indeed *has* both a global minimum and a global maximum (or has either) on this domain. (See, again, Fig. 2.7.) Generally speaking, this will require thinking about (and perhaps graphing) the behavior of $f(x)$ near the (perhaps infinite) boundaries of its domain.

Strategy for finding global extremes

Of course, we have based this strategy on analysis of just a single function f (that of Fig. 2.7). But the strategy does work in considerable generality.

Example 2.6.1 Find the global minimum and the global maximum (when they exist) of $f(x) = 2x^3 + 3x^2 - 36x$ on each of the following domains:

(a) $[-6, 4]$,
(b) $[-6, \infty)$.

Hint: $x^2 + x - 6 = (x + 3)(x - 2)$.

Solution. In each case, the domain is given explicitly, so we need not worry about Step 1 above.

For Step 2, we compute that

$$f'(x) = \frac{d}{dx}\left[2x^3 + 3x^2 - 36x\right] = 6x^2 + 6x - 36.$$

To find the roots of $f'(x)$, we first factor it. We get

$$6x^2 + 6x - 36 = 6(x^2 + x - 6) = 6(x + 3)(x - 2) = 0.$$

This will be zero if and only if one of the factors $x + 3$ or $x - 2$ is zero. In other words, we have found that

$$f'(x) = 0 \text{ precisely when } x = -3 \text{ or } x = 2.$$

Fig. 2.8 The function $f(x)$ of the present example

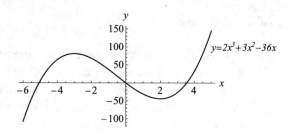

Moreover, $f(x)$ is differentiable everywhere, so there are no points where $f'(x)$ fails to exist. In sum,

the critical points of $f(x)$ are the points $x = -3$ and $x = 2$.

These points are both in each of the domains $[-6, 4]$ and $[-6, \infty)$. We now consider each of those domains separately.

(a) The domain $[-6, 4]$ is closed and bounded, so by the above strategy for finding global extremes, we need only compute, and compare, values of f at the above critical points and at the endpoints of this interval. We calculate that

$$f(-6) = 2 \cdot (-6)^3 + 3 \cdot (-6)^2 - 36 \cdot (-6) = -108,$$
$$f(-3) = 2 \cdot (-3)^3 + 3 \cdot (-3)^2 - 36 \cdot (-3) = 81,$$
$$f(2) = 2 \cdot (2)^3 + 3 \cdot (2)^2 - 36 \cdot (2) = -44,$$
$$f(4) = 2 \cdot (4)^3 + 3 \cdot (4)^2 - 36 \cdot (4) = 32.$$

Comparing these numbers, we see that, on this domain, f attains a global minimum of -108 at the endpoint $x = -6$, and a global maximum of 81 at the critical point $x = -3$.

A sketch of the graph of $f(x)$ helps to illustrate the behavior just elicited, and also helps to illuminate the behavior of f on the next domain under consideration.

(b) Note that the graph of f is rising at $x = 4$. Since there are no critical points to the right of that point, f can only continue to rise as we move to the right of $x = 4$. (In fact, it's not hard to see that $f(x)$ tends to $+\infty$ as x does.) So on the domain $[-6, \infty)$, f has no global maximum (and has the same global minimum as in part (a) of this example) (Fig. 2.8).

In the exercises below, we'll consider additional examples, and will also explore ways in which the *second derivative* f'' can be used to study extremes.

2.6.2   Exercises

2.6.2.1 Part 1: The Derivative and the Shape of a Graph

1. For each of the following, sketch (by hand) a graph of $y = f(x)$ that is continuous on the domain $[-3, 3]$, and is consistent with the given information. On each graph, mark any critical points or extremes, and state whether each of these points is a local minimum, local maximum, or neither.

(a) $f'(x) > 0$ for $x < -1$; $f'(-1)=0$; $f'(x) < 0$ for $-1 < x < 2$; $f'(2) = 0$; $f'(x) > 0$ for $x > 2$.

(b) $f'(x) > 0$ for $x < 2$; $f'(2) = 0$; $f'(x) > 0$ for $x > 2$.

(c) $f'(x) > 0$ for $x < 2$; $f'(2)$ does not exist; $f'(x) < 0$ for $x > 2$.

(d) $f'(x)$ exists for all x, $f'(x) < 0$ for $x < -2$, $f'(x) > 0$ for $-2 < x < 2$, $f(0) = 0$, $f'(x) < 0$ for $x > 2$.

(e) $f'(x) < 0$ for all x except $x = 2$, $f'(2) = 0$.

(f) $f'(x) < 0$ for all x except $x = 2$, the slope of f is infinite at $x = 2$.

(g) $f'(x) > 0$ for $x < -1$, $f'(-1) = 0$, $f'(x) < 0$ for $-1 < x < 2$, $f'(2)$ does not exist, $f'(x) > 0$ for $x > 2$.

2.6.2.2 Part 2: Finding Critical Points

2. For each of the following functions, find all critical points, if any, without using a graphing utility. Assume that the domain of each function is its natural domain—that is, the domain is the set of points where the formula makes sense mathematically.

(a) $f(x) = x^{1/3}$

(b) $f(x) = x^3 + \dfrac{3}{2}x^2 - 6x + 5$

(c) $f(x) = x + \sin x$

(d) $f(x) = \sqrt{1 - x^2}$

(e) $f(x) = 3x^5 - 5x^3$

(f) $f(x) = x^3 10^{-x}$

2.6.2.3 Part 3: Finding Extremes

3. For each of the following graphs (a)–(d), mark any critical points or extremes on the depicted domain. (Assume that each domain is a closed, bounded interval.) Indicate which extremes are local and which are global.

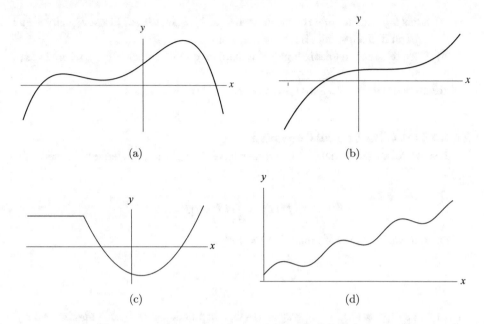

4. Here again are the functions of Exercise 2 above. For each of these functions, use the results of that exercise, together with the above "Strategy for finding global extremes," to find the global minimum and global maximum of each of the given functions on the given closed, bounded interval.

   (a) $f(x) = x^{1/3}$, $[-1, 1]$
   (b) $f(x) = x^3 + \dfrac{3}{2}x^2 - 6x + 5$; $[-3, 3]$
   (c) $f(x) = x + \sin x$; $[-8, 8]$
   (d) $f(x) = \sqrt{1 - x^2}$; $[-1, 1]$
   (e) $f(x) = 3x^5 - 5x^3$; $[-2, 2]$
   (f) $f(x) = x^3 10^{-x}$; $[-2, 2]$

5. (a) Does the function $f(x) = 1/x$ attain a global maximum on the domain $(0, 1]$? If so, where? Please explain. Sketch a graph if it helps, but also explain in words.
   (b) Does this function attain a global minimum on this domain? If so, where? Please explain.

6. Explain why the function $f(x) = 1/x$ does not attain a global minimum on the domain $[1, 2)$.

7. Explain why the function $f(x) = 1/x$ does not attain a global minimum or a global maximum on the domain $(0, \infty)$.

8. (a) Does the function

$$f(x) = \frac{x^2 + 5}{x - 2}$$

attain a global minimum on the interval $(-2, 2)$? If so, where? Please explain. Sketch a graph if it helps, but also explain in words.

(b) Does this function attain a global maximum on this domain? If so, where? Please explain.

9. Repeat the previous exercise for the interval $(2, 6)$.

2.6.2.4 Part 4: The Second Derivative

For these exercises, recall that the *second derivative* f'' of f is the derivative of the derivative of f:

$$f''(x) = \frac{d}{dx}[f'(x)].$$

For example, if $f(x) = x^5$, then $f'(x) = 5x^4$, so

$$f''(x) = \frac{d}{dx}[5x^4] = 20x^3.$$

10. (a) For each of the following graphs (i)–(vi) of a function $y = f(x)$, is *the derivative f'* (not f) increasing or decreasing on the domain shown? (Hint: recall that f' gives the slope of the graph. What does it mean to say this slope is increasing, or decreasing?)

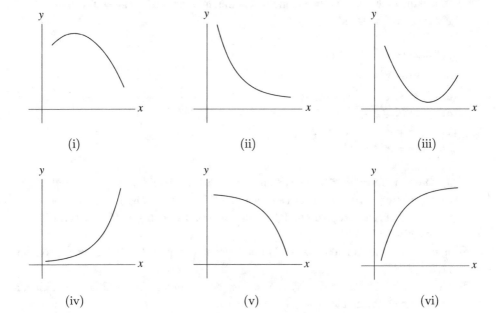

(i)   (ii)   (iii)

(iv)   (v)   (vi)

(b) For each of the above functions, is the second derivative positive or negative on the domain shown? Hint: if the derivative is increasing, what does this say about the derivative of the derivative? Similarly, what do we know about f'' if f' is decreasing?

11. **The geometric meaning of the second derivative.** If f is any function and if $f''(x) > 0$ over some interval, we say the curve is **concave up**, or that it **holds water**, over that interval. If $f''(x) < 0$ over some interval, then we say the curve is **concave down**, or that it **spills water**, over that interval. You should study the graphs in Exercise 10 above until you are clear what this terminology means, and why it makes sense geometrically.

(a) We know that the magnitude of the first derivative tells us how steep the curve is—the greater the magnitude of f', positive or negative, the steeper the curve. What does the magnitude of the second derivative tell us geometrically about the shape of the curve? Complete the sentence: "The greater the magnitude of f'', the _____."

(b) Suppose there are four functions f, g, h, and k, and that $f(0) = g(0) = h(0) = k(0) = 0$, and $f'(0) = g'(0) = h'(0) = k'(0) = 1$. Suppose, moreover, that $f''(0) = 1$, $g''(0) = 5$, $h''(0) = -1$, and $k''(0) = -5$. On the same set of axes, sketch a possible graph of these four functions near the origin.

12. At a point x where the graph of a function f changes concavity from up to down, or vice versa, what must be true about the second derivative $f''(x)$? (Assume that $f''(x)$ exists and is continuous at all points in the domain of f.) Justify your answer.

13. Consider the function $f(x) = 2x^3 + 3x^2 - 36x$ of Fig. 2.8.

(a) By examining the graph visually, estimate: on which interval(s) is the graph of f concave up? On which interval(s) is it concave down?

(b) Check your answer to part (a) of this exercise by computing $f''(x)$, and applying the reasoning of Exercise 10 above.

14. **Second derivative test for maxima and minima.** Explain, in geometric terms, why the following test works.

- If $f'(c) = 0$ and $f''(c) > 0$, then f has a *local minimum* at $x = c$.
- If $f'(c) = 0$ and $f''(c) < 0$, then f has a *local maximum* at $x = c$.

(Hint: think about concavity.)

15. Use the second derivative test to find all local minima and all local maxima of the function $f(x) = x + 2\cos(x)$ on the domain $[0, 4\pi]$. Make sure you specify which of these points are maxima and which are minima. Hint: on $[0, 4\pi]$, the solutions to $\sin(x) = 1/2$ are the points $x = \pi/6$, $x = 5\pi/6$, $x = 13\pi/6$, and $x = 17\pi/6$.

16. (a) Find the (only) critical point of $y = \dfrac{4}{x^2} + x$ on the interval $(0, \infty)$. Explain how you found this value.

 (b) Use the second derivative test to determine whether this point is a local minimum or a local maximum.

 (c) Is this point a global extreme? Explain how you know.

17. Find all critical points of the function $f(x) = x^4 - 42x^2 - 80x$ on the real line. Hint: $4x^3 - 84x - 80 = (x + 4)(x + 1)(x - 5)$. Use the second derivative test to determine which of these points are local minima and which are local maxima.

18. Find all critical points of the function

$$f(x) = \frac{x}{1 + x^2}$$

 on the real line. Use the second derivative test to determine which of these points are local minima and which are local maxima.

19. Notice that the second derivative test says nothing about what happens when $f'(c) = 0$ and $f''(c) = 0$. Show that lots of things can happen in this case, by exhibiting:

 (a) A function f with $f'(0) = f''(0) = 0$ and with a local minimum at $x = 0$;

 (b) A function f with $f'(0) = f''(0) = 0$ and with a local maximum at $x = 0$;

 (c) A function f with $f'(0) = f''(0) = 0$ that has no local extreme point at $x = 0$.

 Hint: think of some very simple polynomial functions.

2.7   Optimization, Part II: Applications

2.7.1   The Problem of the Optimal Soup Can

Suppose you are an aluminum can manufacturer. You need to make a cylindrical can to hold lentil and baby kale soup. The volume V to be contained is specified and fixed, but the proportions—the height h and the radius r—can vary. How can you use the least amount of aluminum to make the can? That is: what proportions will minimize the surface area of the can (Fig. 2.9)?

2.7.2   The Solution

The surface area is the sum of the areas of the two circles at the top and bottom of the can, plus the area of the rectangle that would be obtained if the top and bottom were removed and the side cut vertically and rolled out flat (Fig. 2.10).

 Note that the surface area in question, call it A, depends on r and h:

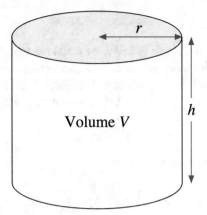

Fig. 2.9 A cylindrical soup can of specified volume V

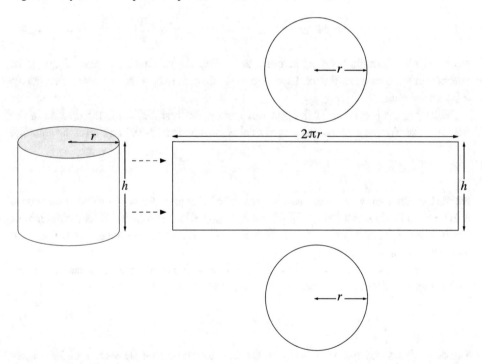

Fig. 2.10 The soup can, deconstructed

$$A = 2\pi r^2 + 2\pi r h. \tag{2.25}$$

Now, were A a function of a *single* variable, we could use techniques and principles from the previous section to find where the global minimum of A, on its domain, occurs. Unfortunately, (2.25) expresses A in terms of the *two* variables r and h.

However, note that r and h are not, in fact, independent of each other. This is because V is fixed, and

$$V = \pi r^2 h.$$

We can solve this for h in terms of r, to get

$$h = \frac{V}{\pi r^2}.$$

Putting this result into (2.25), we may then express A as a function of r alone:

$$A(r) = 2\pi r^2 + 2\pi r \cdot \frac{V}{\pi r^2} = 2\pi r^2 + \frac{2V}{r}. \tag{2.26}$$

Notice that the formula for $A(r)$ involves the *parameter* V. That is, we consider V to be constant in the context of this problem. Our task, then, is to find the value of r that makes $A(r)$ a minimum.

Following the procedure of the previous section, we first determine the domain of the function. Clearly this problem makes physical sense only for $r > 0$. Looking at the equation

$$h = \frac{V}{\pi r^2},$$

we see that although V is fixed, r can, in theory, be as large as we want, provided we choose h to be sufficiently small (which might not be so great for the lentils). Thus the domain of our function $A(r)$ is $0 < r < \infty$, which is not a closed, bounded domain, so we have no guarantee that a minimum exists.

Next we compute $A'(r)$, keeping in mind that the symbols V and π represent constants and that we are differentiating with respect to the variable r. Then

$$A'(r) = 4\pi r - \frac{2V}{r^2}. \tag{2.27}$$

The derivative is undefined at $r = 0$, but this value of r is outside the domain under consideration, so we ignore it.

Next, we set the derivative equal to zero, and solve for any possible critical points:

$$A'(r) = 4\pi r - \frac{2V}{r^2} = \frac{4\pi r^3 - 2V}{r^2} = 0$$

$$0 = 4\pi r^3 - 2V$$

$$4\pi r^3 = 2V$$

$$r = \left(\frac{V}{2\pi}\right)^{1/3}.$$

(We first obtained a common denominator for the terms in $A'(r)$, so we could use the fact that a quotient can only be zero if its numerator equals zero.) Thus $r_{\text{crit}} = \sqrt[3]{V/2\pi}$ is the only critical point.

We can actually sketch the shape of the graph of $A(r)$ versus r, based on this analysis of $A'(r)$. To do this, note that the sign of $A'(r)$ is determined by the numerator in the above formula

$$A'(r) = \frac{4\pi r^3 - 2V}{r^2},$$

since the denominator r^2 is always positive. Moreover,

- When $r < \left(\dfrac{V}{2\pi}\right)^{1/3}$, the numerator is negative, so the graph of $A(r)$ is falling.
- When $r > \left(\dfrac{V}{2\pi}\right)^{1/3}$, the numerator is positive, so the graph is rising.

Note further that

$$\lim_{r \to 0^+} A(r) = \infty \quad \text{and} \quad \lim_{r \to \infty} A(r) = \infty.$$

("$r \to 0^+$" is read "r approaches 0 from the right.") So $A(r)$ looks something like this (Fig. 2.11):

Fig. 2.11 Sketch of the area function $A(r)$

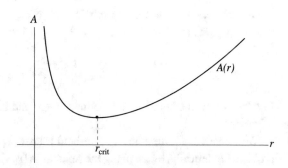

Clearly, then, $A(r)$ has a global minimum at

$$r_{\text{crit}} = \left(\frac{V}{2\pi}\right)^{1/3}.$$

(We could also have seen this using the second derivative test. See Exercise 14 of Sect. 2.6.2.) Further, the minimum area equals

$$A_{\text{crit}} = A(r_{\text{crit}}) = 2\pi(r_{\text{crit}})^2 + \frac{2V}{r_{\text{crit}}} = 2\pi\left(\frac{V}{2\pi}\right)^{2/3} + 2V\left(\frac{2\pi}{V}\right)^{1/3}$$

$$= (2\pi)^{1/3}V^{2/3} + 2(2\pi)^{1/3}V^{2/3}$$

$$= 3(2\pi)^{1/3}V^{2/3}.$$

(It is also clear that there is no *maximum* area, since $A(r)$ tends to $+\infty$ near $r = 0$ and $r = \infty$. Thus, we see again that not *every* optimization problem has a solution.)

Remark 2.7.1 It's not hard to see that, if a function that is continuous on an interval tends to $+\infty$ near either endpoint of that interval, and has only a single critical point on that interval then the global minimum of the function on that interval must occur at that critical point. With this observation in mind, we could have solved the above optimization problem without needing to graph $A(r)$.

2.7.3 Optimization: Some Mathematical Observations

It is interesting to find the value of the height $h = h_{\text{crit}}$ of the above optimal soup can. We have

$$h_{\text{crit}} = \frac{V}{\pi r_{\text{crit}}^2}. \tag{2.28}$$

We'd like to express this more simply in terms of r_{crit}. To do this, let's first solve our formula $r_{\text{crit}} = (V/(2\pi))^{1/3}$ for V, to get $V = 2\pi r_{\text{crit}}^3$. We put this expression for V into Eq. (2.28) for h_{crit}, to get

$$h_{\text{crit}} = \frac{2\pi r_{\text{crit}}^3}{\pi r_{\text{crit}}^2} = 2r_{\text{crit}}.$$

In other words, the height of the optimal (lentil and baby kale) soup can exactly equals its diameter.

We remark that, alternatively, we could have approached our soup can optimization problem numerically, by graphing the function $A(r)$ given by (2.26), and zooming in on the point where $A(r)$ appears, visually, to attain its minimum. But with such an approach, the exact relationship between r_{crit} and h_{crit} (or for that matter, between r_{crit} and V) would not have been so clear and explicit. Moreover, as noted earlier, such an approach would require specifying a numerical value for V, and therefore would only be possible "one V at a time."

Mathematics is sometimes described as the study of *patterns*. Using mathematics—in this case, calculus—to investigate problems can help reveal connecting threads between apparently disparate quantities and phenomena. These connecting threads can help illuminate "what's really going on."

Moreover, the identification of patterns facilitates problem solving. If two problems, in seemingly different contexts, are somehow connected, and we have a solution to one of them, then ideas and techniques from that solution might be applicable to the other problem.

2.7.4 General Strategies for Applied Optimization

From the above soup can problem, we can elicit a general strategy to use in approaching optimization problems. Such a strategy might look like this:

> **Step 1.** Draw a diagram (if relevant). Where possible, label relevant quantities, including variables and parameters.
>
> **Step 2.** Write down a formula for the quantity Q to be optimized, in terms of relevant variables and parameters.
>
> **Step 3.** Use additional information from the problem to find a formula for $Q = Q(x)$ in terms of a single variable x.
>
> **Step 4.** Use calculus—in particular, the principles and techniques of the previous section—to find the desired global minimum or maximum.

Strategy for solving applied optimization problems with calculus

We now apply this strategy to a somewhat different kind of area optimization problem.

Example 2.7.1 A small, trapezoidal vegetable garden is closed off by boards of the same length L on three sides, and by an existing fence (of length at least $3L$) on the other. What is the maximum possible area of this vegetable garden?

Solution. A diagram appears in Fig. 2.12 (you might think of this as an aerial view).

We've attached symbols to the quantities that will be relevant in determining the area in question. Specifically, we see that the enclosed area equals the area of the rectangle in the middle plus the areas of the two triangles on the sides. That is,

$$A = Ly + 2 \cdot \frac{1}{2}xy = Ly + xy = (L + x)y. \tag{2.29}$$

Here, L is a fixed parameter, not a variable. On the other hand, x and y are variables, and may be related to each other by the Pythagorean Theorem, which tells us that $x^2 + y^2 = L^2$.

As it turns out, though, this problem becomes easier to solve if we express both x and y in terms of the angle θ of Fig. 2.12. Specifically, by geometry we have $\cos\theta = x/L$ and $\sin\theta = y/L$, whence

$$x = L\cos\theta; \qquad y = L\sin\theta. \tag{2.30}$$

Then Eq. (2.29) gives us

$$A = A(\theta) = (L + L\cos\theta) \cdot L\sin\theta = L^2(1 + \cos\theta)\sin\theta. \tag{2.31}$$

The domain of $A(\theta)$ can be taken to be the closed, bounded interval $[0, \pi/2]$. (Think about why.) By the "Strategy for finding global extremes" of the previous section, then, our desired global maximum of $A(\theta)$ must occur at $\theta = 0$, $\theta = \pi/2$, or at a critical point of $A(\theta)$ on $[0, \pi/2]$.

To find the critical points, we compute $A'(\theta)$ using Eq. (2.31) and the product rule, and set the result equal to zero:

$$\begin{aligned}
A'(\theta) &= \frac{d}{d\theta}\Big[L^2(1 + \cos\theta)\sin\theta \Big] \\
&= L^2\Big((1 + \cos\theta)\frac{d}{d\theta}\big[\sin\theta \big] + \sin\theta \frac{d}{d\theta}\big[(1 + \cos\theta) \big] \Big) \\
&= L^2\Big((1 + \cos\theta)\cos\theta + \sin\theta(-\sin\theta) \Big) \\
&= L^2\Big(\cos\theta + \cos^2\theta - \sin^2\theta \Big) = 0.
\end{aligned}$$

We divide by L^2, and recall that $\sin^2\theta = 1 - \cos^2\theta$ (see Exercise 17 in Sect. 1.7.4), to get

$$\cos\theta + 2\cos^2\theta - 1 = 0.$$

The left hand side factors; we get

$$(\cos\theta + 1)(2\cos\theta - 1) = 0,$$

so $\cos\theta = -1$ or $\cos\theta = 1/2$. The first of these equations has no solutions on the domain $[0, \pi/2]$; the second has only the solution $\theta = \pi/3$ on this interval.

To determine our global maximum, then, we need only compare the values of $A(\theta)$ at the endpoints $\theta = 0$ and $\theta = \pi/2$, and at the critical point $\theta = \pi/3$. We have:

$$A(0) = L^2(1 + \cos 0)\sin 0 = 0;$$

$$A(\pi/3) = L^2(1 + \cos(\pi/3))\sin(\pi/3) = L^2(1 + 1/2)(\sqrt{3}/2) = \frac{3\sqrt{3}L^2}{4} \approx 1.299L^2;$$

$$A(\pi/2) = L^2(1 + \cos(\pi/2))\sin(\pi/2) = L^2(1 + 0)(1) = L^2.$$

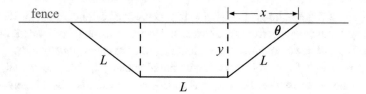

Fig. 2.12 A trapezoidal vegetable garden

The maximum area of $3\sqrt{3}L^2/4$ occurs at $\theta = \pi/3$.

Many of the exercises that follow reveal interesting connections and symmetries among variables, like those observed in our soup can problem above. If you don't observe any such connections or symmetries in a given exercise, reflect on what it is about that exercise that might be making things less symmetric, or "nice," than you might otherwise expect.

2.7.5   Exercises

1. (a) Show that the rectangle of perimeter P whose area is a maximum is a square.
   (b) Use a graphing utility to check your answer to part (a) of this exercise for the special case when $P = 100\,\text{ft}$.
2. (a) Show that the rectangle of area A whose perimeter is a minimum is a square.
   (b) Use a graphing utility to check your answer to part (a) of this exercise for the special case when $A = 100$ square feet.
3. One side of an open field is bounded by a straight river. A farmer has L feet of fencing. How should the farmer proportion a rectangular plot along the river in order to enclose as great an area as possible?
4. An open storage bin with a square base and vertical sides is to be constructed from A square feet of wood. Determine the dimensions of the bin if its volume is to be a maximum. (Neglect the thickness of the wood and any waste in construction.)
5. (a) A roman window is shaped like a rectangle surmounted by a semicircle. If the perimeter of the window is L feet, what are the dimensions of the window of maximum area?
   (b) Use a graphing utility to check your answer for the special case when $L = 100\,\text{ft}$.
6. (a) Suppose the roman window of Exercise 5 has clear glass in its rectangular part and colored glass in its semicircular part. If the colored glass transmits only half as much light per square foot as the clear glass does, what are the dimensions of the window that transmits the most light?
   (b) Use a graphing utility to check your answer for the special case when $L = 100\,\text{ft}$.

7. An open rectangular box is to be made from a piece of cardboard 8 in. wide and 15 in. long, by cutting a square from each corner and bending up the sides. Find the dimensions of the box of largest volume. Use a graphing utility to check your answer.

8. An arrow is fired from ground level, with initial velocity v_0. It's known that, assuming there is no air resistance, the *range R* of this arrow (that is, the horizontal distance traveled before hitting the ground) is given by

$$R = \frac{v_0^2 \sin 2\theta}{g},$$

where θ is the angle at which the arrow is launched, and g is a certain "gravitational constant." (In units of feet per second per second, $g \approx 32$.)

Find the launch angle θ that maximizes the range of the arrow.

9. Find two nonnegative real numbers (not necessarily integers) whose sum is 500 and whose product is as large as possible.

10. Find two nonnegative real numbers (not necessarily integers) whose sum is 500 and whose product is as small as possible.

11. Find two nonnegative real numbers (not necessarily integers) whose product is 500 and whose sum is as small as possible.

12. Do there exist two nonnegative real numbers (not necessarily integers) whose product is 500 and whose sum is as large as possible? If so, what are they? If not, why not?

13. Minimize the surface area of a cylindrical can that has a bottom but *no top*, and has volume V. What are the dimensions (radius and height) of this can, in terms of V? How are these dimensions related, and how are the proportions of this can different from those of the soup can investigated in the section above?

2.8   Summary

The **derivative** is a rate of change. More specifically, the derivative $f'(a)$ represents the **instantaneous rate of change** of the function $y = f(x)$ at the point $x = a$.

The derivative $f'(a)$ tells you "how fast you're going," meaning how fast your output $y = f(x)$ is changing with respect to your input x, at the particular point $x = a$. It may be helpful to think of the derivative as what your speedometer tells you when you view it at a given instant in time. (A speedometer tells you your speed, which is how quickly your displacement, or position, is changing with respect to time.)

There are also **average rates of change**, which signify not how fast you're going at a given instant, but how fast you've gone over the course of your trip. For example, if you travel 110 miles over the course of a two-hour trip, then your average speed over the course of this trip is 110 miles divided by two hours, or 55 miles per hour. More generally, the average rate of change of a function $y = f(x)$, over the course of a "trip" from $x = a$ to

$x = a + \Delta x$, equals the change Δy in y over this trip divided by the change in x. In other words, this average rate of change equals

$$\frac{\Delta y}{\Delta x} = \frac{f(a + \Delta x) - f(a)}{(a + \Delta x) - a} = \frac{f(a + \Delta x) - f(a)}{\Delta x}. \tag{*}$$

If you compute an average rate of change over **shorter and shorter** intervals from $x = a$ to $x = a + \Delta x$, then you should be zeroing in on a pretty good idea of how fast things are changing *at* the point $x = a$. With this in mind, we *define* the instantaneous rate of change, or derivative, $f'(a)$ to be the **limit** of the average rates of change as Δx approaches zero:

$$f'(a) = \lim_{x \to 0} \frac{f(a + \Delta x) - f(a)}{\Delta x}.$$

Geometrically, $(*)$ equals the slope of the **secant line** to the graph of f through the points $(a, f(a))$ and $(a + \Delta x), f(a + \Delta x))$, while $f'(a)$ equals the slope of the **tangent line** to the graph of f at the point $x = a$.

Using a number of formulas for derivatives of specific functions, as well as rules for differentiating "compound" function in terms of their simpler building blocks, we've seen how to compute derivatives in a great variety of situations.

We've also seen how the derivative can be applied to various sorts of optimization problems.

Recall the prediction principle from Sect. 1.1: if you know how fast you're going, then you know how far you'll get in a given amount of time. Since the derivative tells us how fast we're going, we can use it to provide information about how far we'll get, and consequently, where we'll end up. That is, we can use it to **predict**. Earlier we applied this philosophy, in conjunction with the SIR model and Euler's method, to predict the evolution of a disease. In the next chapter, we'll apply the derivative, in similar ways, to prediction in a number of natural contexts.

Differential Equations

3

Rate equations, like the SIR equations from Chap. 1, are often also called **differential equations**. Differential equations are essential tools in many areas of mathematics and the sciences.

In this chapter we explore two related, fundamental aspects of the theory of differential equations:

1. **Mathematical modeling** with differential equations, and
2. **Solving** differential equations, both through numerical techniques like Euler's method and, where possible, through finding formulas that make the equations true.

We introduce two important functions—the **natural exponential function** and the **natural logarithm function**—that are central to many problems that may be modeled by differential equations. And we study dynamical systems (that is, sets of differential equations) modeling a variety of natural phenomena.

3.1 The (Natural) Exponential Function

3.1.1 The Equation $\dfrac{dy}{dt} = ky$

Certain differential equations—in fact, some of the very simplest—arise over and over again in a remarkable variety of contexts. The functions to which they give rise are among the most important in mathematics.

One of the most basic differential equations is

© The Author(s), under exclusive license to Springer Nature Switzerland AG 2023   125
E. Stade and E. Stade, *Calculus: A Modeling and Computational Thinking Approach*,
Synthesis Lectures on Mathematics & Statistics,
https://doi.org/10.1007/978-3-031-24681-4_3

$$\frac{dy}{dt} = ky \quad \text{(where } k \text{ is a constant).} \tag{3.1}$$

It is also one of the most useful, because *it describes a quantity y that is proportional to its rate of change*. Many real-world phenomena evolve in this way, at least roughly. For example: in Sect. 1.7, we used this differential equation to model population growth, as well as (in the Exercises for that section) bacterial growth, radioactive decay, and Moore's Law for transistors on a microchip. Later in this chapter we will use it to model other phenomena, such as the dissolving of substances water, and how radiation penetrates solid objects.

We've already encountered certain solutions to the differential equation (3.1). Specifically: recall from Sect. 2.3 that, if $y = b^t$, then

$$\frac{dy}{dt} = \ln(b)b^t, \tag{3.2}$$

where

$$\ln(b) = \lim_{\Delta x \to 0} \frac{b^{\Delta x} - 1}{\Delta x}. \tag{3.3}$$

But note that the quantity b^t on the right-hand side of (3.2) is what we called y in the first place. So (3.2) reads

$$\frac{dy}{dt} = \ln(b)y. \tag{3.4}$$

In other words, $y = b^t$ *is* a solution to the differential equation (3.1), in the case where $k = \ln(b)$.

To study these solutions more systematically, it will be convenient, first, to consider more closely the case $k = 1$.

3.1.2　The Equation $\dfrac{dy}{dt} = y$, and the Natural Exponential Function

In Sect. 2.3 we noted that, by Eq. (3.3) above and the meaning of limits, we have

$$\ln(b) \approx \frac{b^{\Delta x} - 1}{\Delta x} \tag{3.5}$$

for Δx small. We used this observation, with $\Delta x = 0.000001$, to find that $\ln(2) \approx 0.69314$. Of course, this number is less than one. A similar approximation shows that $\ln(3) \approx 1.09861$, which is, of course, greater than one. So we might expect that, somewhere between 2 and 3, there is a number b such that $\ln(b)$ exactly *equals* one.

We would be correct! We'll justify this shortly, but first, taking on faith for the moment that there *is* such a base b, and only one, let's give this base a name: let's call it e.

Definition 3.1.1 The number e is the unique real number satisfying $\ln(e) = 1$.

Some important properties of the number e are as follows.

1. Since $d[b^t]/dt = \ln(b)b^t$ for any positive base b (as noted above), and since $\ln(e) = 1$, we have

$$\frac{d}{dt}[e^t] = \ln(e)e^t = 1 \cdot e^t = e^t.$$

The derivative of e^t is e^t

In other words, **the natural exponential function $y = e^t$ is equal to its own derivative**. This property of the function $y = e^t$ is largely what makes the base e so useful and ubiquitous. (The *raison d'etre* of the base e is a calculus thing!)

In fact, the natural exponential function is *so* natural that we typically call it, simply, **the exponential function** (to the chagrin, perhaps, of $y = b^t$ for other values of b).

Further, as we'll soon see, solutions to the differential equation (3.1) may be written explicitly and simply in terms of t, k, and e.

2. Putting $b = e$ into Eq. (3.5) above, and recalling that $\ln(e) = 1$, we get

$$1 \approx \frac{e^{\Delta x} - 1}{\Delta x} \tag{3.6}$$

for Δx small. We use this equation to approximate e, as follows. We multiply both sides of (3.6) by Δx, add 1 to both sides of our result, and then raise both sides to the power of $1/\Delta x$. We find that

$$e \approx (1 + \Delta x)^{1/\Delta x} \tag{3.7}$$

for Δx small.

For example, using $\Delta x = 0.000001$ gives

$$e \approx (1.000001)^{1,000,000} = 2.71828\ldots.$$

And we could get a better approximation with a smaller Δx. (It turns out that e is irrational, so its decimal expansion has infinitely many places.)

The fact (which we won't prove) that the right-hand side of (3.7) *has* a limit as $\Delta x \to 0$ is what tells us that the base e exists, and it's not hard to show that this base e is the only one whose natural logarithm equals one.

The number e is, like π, one of the most important and ubiquitous in mathematics.

Here are some examples involving differentiation of the (natural) exponential function.

Example 3.1.1 (a) Find:

(a) $\dfrac{d}{dx}[e^{ax}]$ (where a is a constant);   (b) $\dfrac{d}{dz}[e^{3\sin(z)}]$;   (c) $\dfrac{d}{dz}[3\sin(e^{z})]$.

Solution. (a) The derivative of e^{x}, with respect to x, is e^{x}, so by the chain rule, the derivative of e to a function of x, with respect to x, is e to that same function of x, times the derivative of that function of x. So

$$\frac{d}{dx}[e^{ax}] = e^{ax}\frac{d}{dx}[ax] = e^{ax}\cdot a = ae^{ax}.$$

(b) By similar reasoning,

$$\frac{d}{dz}[e^{3\sin(z)}] = e^{3\sin(z)}\frac{d}{dz}[3\sin(z)] = e^{3\sin(z)}\cdot 3\cos(z) = 3\cos(z)e^{3\sin(z)}.$$

(c) The derivative of the sine function is the cosine function, so by the chain rule,

$$\frac{d}{dz}[3\sin(e^{z})] = 3\frac{d}{dz}[\sin(e^{z})] = 3\cos(e^{z})\frac{d}{dz}[e^{z}] = 3\cos(e^{z})e^{z}.$$

3.1.3   The Equation $\dfrac{dy}{dt} = ky$, Again

In Example 3.1.1(a) above, we differentiated e to a constant times the independent variable. Notice that the result of this differentiation was that constant times the function we started with!

More generally, consider the function

$$y = Ce^{kt},$$

where both C and k are constants, and $k > 0$. We claim that y has two important features: (i) the rate of growth of y is proportional to y, with constant of proportionality k; and (ii) initially, meaning when $t = 0$, y is equal to the contant C.

We prove these claims as follows: first, by the chain rule,

$$\frac{dy}{dt} = \frac{d}{dt}[Ce^{kt}] = C\frac{d}{dt}[e^{kt}] = Ce^{kt}\frac{d}{dt}[kt] = Ce^{kt}\cdot k = y\cdot k = ky.$$

And second,

$$y(0) = Ce^{k\cdot 0} = Ce^{0} = C\cdot 1 = C$$

(since any positive number to the 0th power is one).

We summarize the above in the following.

> $y = Ce^{kt}$ **is the solution to the "initial value problem"**
> $$\frac{dy}{dt} = ky; \qquad y(0) = C.$$
> **Here k and C are constants, and $k > 0$.**

Solution to the "exponential growth" initial value problem

A couple of remarks are in order.

- By "initial value problem," we mean one or more differential equations, together with one or more *initial conditions*, meaning specification of how things look at some particular point in time (often at $t = 0$). In the exponential growth initial value problem, the equation $dy/dt = ky$ is of course the differential equation, and $y(0) = C$ is the initial condition.
- We have shown that $y = Ce^{kt}$ is **a** solution to the exponential growth initial value problem, but here, we are saying more. We are saying it is **the** solution, meaning there are no others. We have not proved this, but it follows from general results in the theory of differential equations, and we will take it on faith.

Example 3.1.2 A population P grows at a yearly rate equal to 0.3 times the population size. Assume that $P = 100,000$ when $t = 0$.

(a) Write down the initial value problem for P.
(b) Write down a formula for $P(t)$.
(c) What is the population after four years?

Solution. (a) We have
$$\frac{dP}{dt} = 0.3P; \qquad P(0) = 100,000.$$

(b) By the above-stated results concerning the solution to the exponential growth initial value problem,
$$P(t) = 100,000e^{0.3t}.$$
Here, t is in years and P is in individuals.

(c) $P(4) = 100,000e^{0.3 \cdot 4} = 332,012$ individuals (to the nearest whole number of individuals).

In part (c) of the above example, we found the value of $P(t)$ corresponding to a particular value of t. In a later section, we'll see how to do an "inverse" process: we'll see how to find the value of t that corresponds to a particular value of $P(t)$. That is, we'll answer a "how long" question, as opposed to a "how large" question. (We'll solve for the independent variable instead of the dependent variable.)

We now note that, strictly speaking, $y = Ce^{kt}$ satisfies the initial value problem given by $dy/dt = ky$ and $y(0) = C$ for *any* real number k, not just when k is positive. However, if k is negative, then the equation $dy/dt = ky$ tells us that y has a negative derivative (assuming y itself is positive), so that y is decreasing. In this case, we have *decay* rather than growth, and it's a bit jarring to still refer to the "exponential growth" initial value problem. So we typically formulate decay situations a bit differently. Namely, we still require that k be positive, but we write $-k$ instead of k for our constant of proportionality. Our solution, then, also entails a factor of $-k$ instead of k, and we have the following.

$y = Ce^{-kt}$ **is the solution to the initial value problem**
$$\frac{dy}{dt} = -ky; \qquad y(0) = C.$$
Here k and C are constants, and $k > 0$.

Solution to the "exponential decay" initial value problem

Example 3.1.3 Sugar dissolves in water at a rate proportional to the amount present. Write an equation for the amount $S(t)$ of sugar remaining after t minutes, in terms of a "per unit decay rate" k and an initial amount S_0. Assuming that $S(t)$ is measured in pounds, what are the units for k?

Solution. We're given that $S'(t) = -kS(t)$ (the minus sign indicates decay), and that $S(0) = S_0$. So, by the above-stated results concerning the exponential decay initial value problem,

$$S(t) = S_0 e^{-kt}.$$

The units for k are given by the original differential equation: if $S'(t) = -kS$ and $S(t)$ is measured in pounds, and t is in minutes, then k must be in minutes^{-1}, so that units match up on the two sides of this differential equation.

In a later section, we'll see how to find a numerical value for the parameter k, given certain additional information (such as, for example, the amount of salt remaining five minutes later).

3.1.4 Basic Properties of the (Natural) Exponential Function

The function $y = e^x$ has a number of useful properties. Among these are the following.

Proposition 3.1.1

(i) $e^0 = 1$.

(ii) $e^x > 0$ *for all real numbers* x.

(iii) $e^{-x} = 1/e^x$ *for all real numbers* x.

(iv) $e^{x+y} = e^x e^y$ *for all real numbers* x *and* y.

(v) $e^{x-y} = e^x/e^y$ *for all real numbers* x *and* y.

(vi) $(e^r)^s = e^{rs}$ *for all real numbers* r *and* s.

(vii) $\frac{d}{dx}[e^x] = e^x$.

(viii) $\lim_{x \to +\infty} e^x = +\infty$.

(ix) $\lim_{x \to -\infty} e^x = 0$.

(x) *The graph of* $y = e^x$ *looks like* $Fig. 3.1$.

The first six of these properties are still true if "e" is replaced by "b," where b is any positive number. The seventh property is unique to the base e, though a slight modification holds for other bases. (What modification is required?)

The eighth property tells us that e^x increases without bound, or "blows up" (in the positive, rather than the negative, vertical direction), as we go further out along the positive x axis. The ninth property tells us that e^x gets arbitrarily close to zero as we go further out along the *negative* x axis.

Properties (viii) and (ix) remain true, and the graph of property (x) retains the same basic shape (though a different vertical scale) if e is replaced by b, **as long as b is larger than 1**. If $b \leq 1$, then these properties require modification. See the exercises below.

Example 3.1.4 Find $\dfrac{d}{dx}[e^x e^{7-x}]$ in two ways:

(a) Differentiate and then simplify;
(b) Simplify and then differentiate.

Fig. 3.1 The natural exponential function $y = e^x$

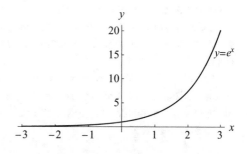

Solution. (a) By the product and chain rules,

$$\frac{d}{dx}[e^x e^{7-x}] = e^x \frac{d}{dx}[e^{7-x}] + e^{7-x}\frac{d}{dx}[e^x] = e^x e^{7-x}\frac{d}{dx}[7-x] + e^{7-x}e^x$$
$$= e^x e^{7-x}(-1) + e^{7-x}e^x = 0,$$

since the two summands cancel.

(b) By the above property (iv) of the natural exponential function, we have $e^x e^{7-x} = e^{x+7-x} = e^7$, which is just a *constant*. So

$$\frac{d}{dx}[e^x e^{7-x}] = \frac{d}{dx}[e^7] = 0$$

(again).

Remark 3.1.1 By Proposition 3.1.1 above, the function $y = e^x$ grows without bound as x gets large. But one can say more: one can say that, as $x \to \infty$, the function $y = e^x$ grows faster than $y = x^2$, faster than $y = x^5$, faster than $y = x^{273,000}$—**faster than any power x^p of x** (no matter how large p is)!

More specifically: in Exercise 5 below, we outline a demonstration of the fact that

$$\lim_{x \to +\infty} \frac{x^p}{e^x} = 0$$

for any real number p (no matter how large). In this sense, the growth of e^x eventually "wins out" over that of x^p, regardless of how large p may be.

The speed of exponential growth has had an impact in computer science. In many cases, the number of operations needed to calculate a particular quantity is a power of the number of digits of precision required in the answer. Sometimes, though, the number of operations is an *exponential* function of the number of digits. When that happens, the number of operations can quickly exceed the capacity of the computer. In this way, some problems that can be solved by an algorithm that is straightforward in theoretical terms are intractable in practical terms.

Any quantity subject to exponential growth grows "very quickly," in the sense made precise in Remark 3.1.1 above. Thus, exponential growth can have serious ramifications for phenomena like population growth and various others. A number of these are explored in the coming sections.

3.1.5 Exercises

3.1.5.1 Part 1: The Functions b^x

1. What are the analogs of properties (viii) and (ix) of the exponential function above, if we replace e with the base $b = 1$? That is, what are

$$\lim_{x \to +\infty} 1^x \quad \text{and} \quad \lim_{x \to -\infty} 1^x?$$

2. (a) Use a graphing utility to plot $y = 3^x$ and $y = (1/3)^x$ on the same axes. Describe how these two functions compare, graphically.

   (b) Repeat part (a) of this exercise for $y = 7^x$ and $y = (1/7)^x$.

   (c) We noted above that properties (viii) and (ix) of the function $y = e^x$ still hold if we replace e by b, as long as $b > 1$. In light of parts (a) and (b) of this exercise, what do you think the analogous properties are for $b < 1$? Hint: recall that $b^{-x} = 1/b^x$.

3.1.5.2 Part 2: Differentiating Exponential Functions

3. Differentiate the following functions.

   (a) $f(x) = 7e^{3x}$

   (b) $y = Ce^{kx}$, where C and k are constants.

   (c) $g(t) = 1.5e^t$

   (d) $q = 1.5e^{2t}$

   (e) $r(x) = 2e^{3x} - 3e^{2x}$

   (f) $z(t) = e^{\cos t}$

   (g) $y = x^4 e^{4x}$

   (h) $f(v) = \dfrac{e^v}{e^v + 1}$

   (i) $y = \tan(x^5 e^{5x})$

   (j) $y = e^{x^5 \tan 5x}$

   (k) $y = e^{e^x}$ (Note: this means $e^{(e^x)}$, not $(e^e)^x$.)

   (l) $q = (e^\pi)^\pi$

3.1.5.3 Part 3: Powers of e

4. Rewrite each of the following quantities as a whole number, or as e to a single expression. For each, please state which of the above properties (i)–(vi) of the exponential function you have used. The first one has been done for you, to illustrate what is meant.

   (a) $(e^x)^2/e^y$ (**Solution:** $(e^x)^2/e^y = e^{2x}/e^y = e^{2x-y}$, by properties (vi) and (v).)

   (b) $\dfrac{e^3 e^2}{e^5}$

(c) $e^{x^2}/(e^x)^2$ (Note: e^{x^2} means $e^{(x^2)}$, not $(e^x)^2$.)

(d) $\left((e^2 e^3 e^4)/e\right)/e^3$

(e) $\left(e^{y^2-5y}\right)^{1/y} e^5 / e^y$

(f) $\left((e^3)^y (e^{-y})^3 / e^x\right)^2$

3.1.5.4 Part 4: The Growth of e^x

We claimed in the section above that

$$\lim_{x\to\infty} \frac{x^p}{e^x} = 0$$

for any real number p. In other words, e^x grows faster than any power of x. (This is clear for negative powers of x, since these don't grow at all—in fact, they decay—as $x \to \infty$. It's also clear for the zeroth power of x, since this is constant. But it's not so obvious for positive powers.) Here we outline a proof of this fact.

Let p be a given real number, and define

$$f(x) = \frac{x^p}{e^x} = x^p e^{-x}.$$

Our goal is to show that

$$\lim_{x\to\infty} f(x) = 0.$$

(We'll assume throughout that $x > 0$.)

5. (a) Explain why $f(x) > 0$ for all positive numbers x.

   (b) Use the product rule and some algebra to show that

   $$f'(x) = x^{p-1} e^{-x}(p - x).$$

   (c) Using part (b) of this exercise, explain why f is decreasing for $x > p$.

   By parts (a) and (c) of this exercise, f is steadily decreasing for $x > p$, but never dips below zero. It stands to reason, then (and in fact follows from a property of the real line called the **greatest lower bound property**), that the graph of f must "level off," as x increases, at some nonnegative height L.

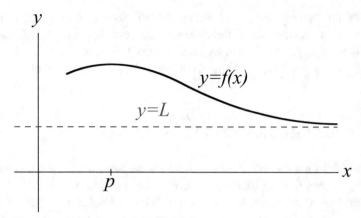

If we can show that $L = 0$, then we'll be done; we will have shown that $f(x)$ "levels off" at $y = 0$ as x gets large. In other words, we will have shown that

$$\lim_{x \to \infty} f(x) = 0.$$

So let us show that $L = 0$.

(d) Use part (b) of this exercise and the definition of $f(x)$ to show that

$$f(x) = \frac{x}{p - x} \cdot f'(x).$$

(e) Noting that

$$\frac{x}{p - x} = \frac{x}{p - x} \cdot \frac{1/x}{1/x} = \frac{1}{\frac{p}{x} - 1},$$

what happens to $\dfrac{x}{p - x}$ as $x \to \infty$?

(f) From the above picture, what must

$$\lim_{x \to \infty} f'(x)$$

be equal to? (Recall that the derivative measures slope.)

(g) Use parts (d) (e), and (f) of this exercise to show that

$$\lim_{x \to \infty} f(x) = 0,$$

which is exactly what we wanted to show.

6. Show that e^{-x} approaches zero, as $x \to \infty$, faster than any power of x. In other words, show that

$$\lim_{x \to \infty} \frac{e^{-x}}{x^q} = 0.$$

for any real number q. (Of course, if $q \geq 0$, then x^q does not approach zero *at all* as $x \to \infty$. But the point of this exercise is that, even if q is a (perhaps very large) negative number, so that x^q does indeed tend to zero $x \to \infty$, e^{-x} does so faster.)
Hint: rewrite e^{-x}/x^q, and use the result of the previous exercise.

3.1.5.5 Part 5: Solving $y' = ky$ Using e^t
7. **Poland**. Refer to Example 1.7.3 in Sect. 1.7 above.

(a) Write out the initial value problem for the population P given in that example.
(b) Write a formula for the solution of this initial value problem.
(c) Use your formula from part (b) to find the population of Poland in the year 2005, to the nearest whole person. According to this formula, what was the population in 1965? In 2020?

8. **Bacterial growth**. Refer to Exercise 22 of Sect. 1.7.4.

(a) Assuming that we begin with the colony of bacteria weighing 32 g, write out the initial value problem that summarizes the information about the weight P of the colony.
(b) Write a formula for the solution P of this initial value problem.
(c) How much does the colony weigh after 30 min? after 2 h?

9. **Radioactivity**. Refer to Exercise 23 of Sect. 1.7.4.

(a) Assuming that, when we begin, the sample of radium weighs 1 g, write out the initial value problem that summarizes the information about the weight R of the sample.
(b) How much did the sample weigh 20 years ago? How much will it weigh 200 years hence?

10. **Intensity of radiation**. As gamma rays travel through an object, their intensity I decreases with the distance x that they have travelled. This is called **absorption**. The absorption rate dI/dx is proportional to the intensity. For some materials the multiplier in this proportion is large; they are used as radiation shields.

(a) Assume that the unshielded intensity of the gamma rays (that is, the intensity when $x = 0$) is I_0. Write down an initial value problem that models the intensity of gamma rays $I(x)$ as a function of distance x.
(b) What quantity or quantities in your above initial value problem reflect(s) how well your material shields against radiation?
(c) Write a formula for the intensity I in terms of the distance x.

11. In this exercise, you will find a solution for the initial value problem $y' = ky$ and $y(t_0) = C$. (Notice that this isn't the original initial value problem considered earlier in this section, because t_0 was 0 originally.)

   (a) Show that the function $y = Ae^{kt}$ satisfies the differential equation in question, for any constant A.
   (b) Find A in terms of k, C and t_0, by considering the given initial condition.

3.1.5.6 Part 6: More Modeling with Exponentials

12.

   (a) **Newton's law of cooling.** Verify that

   $$C(t) = 70e^{-0.1t} + 20$$

   is a solution to the initial value problem

   $$C'(t) = -0.1(C - 20), \qquad C(0) = 90.$$

   (b) How is part (a) of this exercise related to Exercise 24 of Sect. 1.7?
   (c) Verify that
   $$C(t) = (C_0 - A)e^{-kt} + A$$

   is a solution to the the initial value problem

   $$C'(t) = -k(C - A), \qquad C(0) = C_0.$$

   (d) Referring to the setting described in Exercise 24 of Sect. 1.7, describe a scenario that might be modeled by the function $C(t)$ of part (c) of this exercise.

13. In *An Essay on the Principle of Population*, written in 1798, the British economist Thomas Robert Malthus (1766–1834) argued that food supplies grow at a constant rate, while human populations naturally grow at a constant *per capita* rate. He therefore predicted that human populations would inevitably run out of food (unless population growth was suppressed by unnatural means).

   (a) Write differential equations for the size P of a human population and the size F of the food supply that reflect Malthus' assumptions about growth rates.
   (b) Keep track of the population in millions, and measure the food supply in millions of units, where one unit of food feeds one person for one year. Malthus' data suggested to him that the food supply in Great Britain was growing at about 0.28 million units per year, and the per capita growth rate of the population was 2.8% per year. Let $t = 0$ be the year 1798, when Malthus estimated the population of the British Isles

was $P = 7$ million people. He assumed his countrymen were on average adequately nourished, so he estimated that the food supply was $F = 7$ million units of food. Using these values, write formulas for the solutions $P = P(t)$ and $F = F(t)$ of the differential equations in (a).

(c) Use the formulas in (b) to calculate the amount of food and the population at 25 year intervals for 100 years. Use these values to help you sketch graphs of $P = P(t)$ and $F = F(t)$ on the same axes.

(d) The per capita food supply in any year equals the ratio $F(t)/P(t)$. What happens to this ratio as t grows larger and larger? (Use your graphs in (c) to assist your explanation. You can also use results from Exercise 5 of this section, above, if you completed that exercise.) Do your results support Malthus's prediction? Explain.

3.2   The Natural Logarithm Function

We further examine some exponential growth/decay contexts considered in Sect. 3.1, and develop some tools to solve certain kinds of problems that arise in those contexts.

3.2.1   Solving the Equation $e^a = b$ for a

In Sect. 3.1, we solved exponential growth and decay problems problems of the "how large/how much" variety. That is: using formulas like

$$P = P_0 e^{kt} \quad \text{and} \quad R = R_0 e^{-kt}, \tag{3.8}$$

we were able to answer some questions like "how large is this population after that many years," or "how much of this radioactive substance is left after that many hours?" Answers to such problems entail simply putting the appropriate value of t into the appropriate formula (3.8) for the exponentially growing/decaying substance.

What we have not yet considered are answers to "how long" questions, in exponential growth/decay situations. That is, how do we solve equations like (3.8) for t, given values of the other variables and constants involved?

For example, consider the following.

Example 3.2.1 Suppose a population, initially comprising 100,000 persons, is growing at the per capita rate of $k = 3$ births per thousand persons per year.

(a) Write down an initial value problem modeling this situation.
(b) How large will this population be 37 years from now?
(c) How long will it take the population to double?

Solution. (a) Denoting time in months by t, and population in individuals by $P(t)$, we have the initial value problem

$$P'(t) = 0.003 P(t); \qquad P(0) = 100,000.$$

(b) Using the results of Sect. 3.1, we know that the solution to the problem is the exponential function

$$P(t) = 100,000 \, e^{0.003t}.$$

The size of the population 37 years from now will therefore be

$$P(37) = 100,000 \, e^{0.003(37)}$$
$$= 100,000 \cdot 1.117395$$
$$\approx 111,740 \text{ people.}$$

(c) To find out by when the population will double, we want to find a value of t such that $P(t) = 200,000$. In other words, we need to solve for t in the equation

$$100,000 \, e^{0.003\,t} = 200,000.$$

Dividing both sides by 100,000, we have

$$e^{0.003t} = 2. \tag{3.9}$$

At the moment, we don't have the technology to solve (3.9), because one side is in exponential form, but the other isn't.

To solve (3.9), we need answer this question: how do we "get the 0.003t out of the exponent"? The answer lies, perhaps surprisingly, in the quantity $\ln(b)$ that arises when we differentiate $f(x) = b^x$ (see Example 2.3.2). Specifically, the following is true:

$$\boxed{\ln(e^a) = a \textbf{ for any real number } a.} \tag{3.10}$$

We will use Eq. (3.10) to solve Example 3.2.1c. But first, we wish to prove (3.10).

The proof, while perhaps a bit unintuitive, is a nice illustration of the power of calculus. It goes like this: first we give e^a a name; let's call it b. Since

$$\frac{d}{dx}[b^x] = \ln(b)b^x$$

for any positive number b, and since e^a is a positive number for any real number a, we have

$$\frac{d}{dx}\left[(e^a)^x\right] = \ln(e^a)(e^a)^x. \tag{3.11}$$

Since $(e^a)^x = e^{ax}$ by Proposition 3.1.1(vi), Eq. (3.11) reads

$$\frac{d}{dx}\left[e^{ax}\right] = \ln(e^a)e^{ax}. \tag{3.12}$$

The left hand side equals ae^{ax} by Example 3.1.1(a) above, so (3.12) yields

$$ae^{ax} = \ln(e^a)e^{ax}. \tag{3.13}$$

Divide both sides by e^{ax} to get $\ln(e^a) = a$, as claimed.

Let's return to Example 3.2.1(c). There we encountered the question that led to the introduction of the logarithm function in the first place, namely: how do we solve the equation

$$e^{0.003t} = 2?$$

We now have a way. Specifically, we take the natural logarithm on both sides of this equation, to get

$$\ln(e^{0.003t}) = \ln(2).$$

But by Eq. (3.10) above, $\ln(e^{0.003t}) = 0.003t$. So we get

$$0.003t = \ln(2)$$

or, dividing by t and using a calculator,

$$t = \frac{\ln(2)}{0.003} = 231.049....$$

So, in answer to the original question: it takes a little over 231 years for this population to double.

Returning to natural logarithms and exponentials in general, we note next that there is a counterpart to Eq. (3.10)—that is, an equation that gives us a simple expression for $e^{\ln(b)}$, for any positive number b. To see this, we take the natural exponential function of both sides of (3.10), to get

$$e^{\ln(e^a)} = e^a. \tag{3.14}$$

Let's give a name to e^a; let's call it b. Putting the name b in for e^a in both sides of (3.14) gives

$$e^{\ln(b)} = b. \tag{3.15}$$

Now a can be anything, so $b = e^a$ can be anything *positive*. (As a ranges over all real numbers, e^a ranges over all positive numbers, as we see from the graph of $y = e^x$ above.) So (3.15) holds for all positive b. In sum,

$$\boxed{e^{\ln(b)} = b \text{ for any positive number } b.} \tag{3.16}$$

This is our desired counterpart to (3.10).

The Eqs. (3.10) and (3.16) together express the fact that **the natural logarithm and natural exponential functions are inverses of each other**. Here, the term "inverse" is used not to mean that these functions are reciprocals of each other (they're not!), but instead that they "undo" each other. That is: if you start with a positive number b, take its natural logarithm, and then raise e to the result, then (by (3.16)) you get back what you started with. Similarly: if you start with a real number a, raise e to that number, and then take the natural logarithm of the result, then (by (3.10)) you get back what you started with.

Many pairs of functions that share a key on a calculator—tangent and arctangent, square root and squaring—are inverses of each other (on appropriate domains). There are even functions that are their own inverses—apply such a function to any number, then apply this same function to the result, and you're back at the original number. (Consider $f(x) = 1 - x$. Can you think of others?) We will say more about inverse functions in the next section.

3.2.2   Properties of the Natural Logarithm Function

Algebraic properties. The inverse relationship between exponents and logarithms—that is, the fact that they "undo" each other—allows us to translate each property of the exponential function into a corresponding statement about the logarithm function. We list the major pairs of properties below.

exponentialversion	logarithmicversion
$e^0 = 1$	$\ln(1) = 0$
$e^{a+b} = e^a e^b$	$\ln(mn) = \ln(m) + \ln(n)$ $(m, n > 0)$
$e^{a-b} = e^a / e^b$	$\ln(m/n) = \ln(m) - \ln(n)$ $(m, n > 0)$
$(e^a)^s = e^{as}$	$\ln(m^s) = s \cdot \ln(m)$ $(m > 0)$
range of e^x is all positive reals	domain of $\ln(x)$ is all positive reals
domain of e^x is all real numbers	range of $\ln(x)$ is all real numbers
$e^x \to 0$ as $x \to -\infty$	$\ln(x) \to -\infty$ as $x \to 0$
e^x goes to $+\infty$ faster than x^p, for any $p > 0$	$\ln(x)$ goes to $+\infty$ slower than $x^{1/p}$, for any $p > 0$

For each of these pairs of properties, we can use the exponential property and the inverse relationship between exponential and logarithmic functions to establish the corresponding logarithmic property. As an example, we will establish the second property. You might want to think about how to demonstrate the others. (See Exercise 3 below.)

Proof of the second property. We wish to use the property $e^{a+b} = e^a e^b$ to deduce the property $\ln(mn) = \ln(m) + \ln(n)$. To do this note that, since we are assuming m and n to be positive numbers, there must be real numbers a and b such that $e^a = m$ and $e^b = n$. But then

$$\ln(mn) = \ln(e^a e^b) = \ln(e^{a+b}) = a + b = \ln(m) + \ln(n),$$

and our proof is complete.

Example 3.2.2 Simplify:

$$\ln\left(\frac{a^{10}e^{x^2}}{b^{\sin(x)}}\right).$$

Solution. By some of the properties listed above (which ones?),

$$\ln\left(\frac{a^{10}e^{x^2}}{b^{\sin(x)}}\right) = \ln\left(a^{10}e^{x^2}\right) - \ln\left(b^{\sin(x)}\right)$$

$$= \ln\left(a^{10}\right) + \ln\left(e^{x^2}\right) - \ln\left(b^{\sin(x)}\right)$$

$$= 10\ln(a) + x^2 - \sin(x)\ln(b).$$

Geometric properties. We now consider the function $y = \ln(x)$ geometrically: what does the graph of this function look like?

To answer, we begin by noting that (obviously) the function $y = e^x$ takes an input a to an output e^a. But now observe that, by (3.10), $\ln(e^a) = a$, so that the function $y = \ln(x)$ takes an input e^a to an output a. Schematically, we have this picture:

$$a \longrightarrow \boxed{\begin{array}{c} \text{the function} \\ y = e^x \end{array}} \longrightarrow e^a$$

$$e^a \longrightarrow \boxed{\begin{array}{c} \text{the function} \\ y = \ln(x) \end{array}} \longrightarrow a$$

In other words: *the function $y = \ln(x)$ takes the function $y = e^x$, and* **reverses** *the roles of input and output* (Fig. 3.2).

Geometrically, swapping input with output amounts to swapping the horizontal with the vertical. And as one can show, this is the same as *reflecting* everything about the line $y = x$. CONCLUSION: the graph of $y = \ln(x)$ **is** the graph of $y = e^x$, reflected about the line $y = x$!

3.2.3   The Derivative of the Logarithm Function

Because the functions $y = e^x$ and $y = \ln(x)$ "undo" each other, our knowledge about the derivative of the first of these functions can be transformed into a formula for the derivative of the second.

Here's how. We begin with Eq. (3.16), with x in place of b:

$$e^{\ln(x)} = x.$$

Fig. 3.2 The graph of
$y = \ln(x)$ (in green) is the
graph of $y = e^x$ (in red),
reflected about the line $y = x$
(dashed)

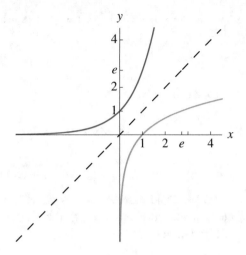

We differentiate both sides:

$$\frac{d}{dx}\left[e^{\ln(x)}\right] = \frac{d}{dx}[x]. \tag{3.17}$$

The right-hand side is simply equal to 1. The left-hand side equals $e^{\ln(x)} \cdot d[\ln(x)]/dx$, by
the chain rule. So (3.17) gives

$$e^{\ln(x)}\frac{d}{dx}[\ln(x)] = 1$$

or, dividing both sides by $e^{\ln(x)}$,

$$\frac{d}{dx}[\ln(x)] = \frac{1}{e^{\ln(x)}}. \tag{3.18}$$

We've found a formula for the derivative of $\ln(x)$! And we can further simplify this formula
by plugging into the right-hand side the fact that, again, $e^{\ln(x)} = x$. So (3.18) reads

$$\boxed{\frac{d}{dx}[\ln(x)] = \frac{1}{x}}$$

Derivative of the natural logarithm function

Example 3.2.3 Find: (a) $\dfrac{d}{dx}[\cos(1 - \ln(x))]$;   (b) $\dfrac{d}{dx}[\ln(1 - \cos(x))]$;   (c) $\dfrac{d}{dx}\left[\ln\left(\dfrac{x^2}{\ln(x^2)}\right)\right]$.

Solution. (a) By the chain rule,

$$\frac{d}{dx}[\cos(1-\ln(x))] = -\sin(1-\ln(x))\frac{d}{dx}[1-\ln(x)]$$

$$= -\sin(1-\ln(x))\left(0-\frac{1}{x}\right) = \frac{\sin(1-\ln(x))}{x}.$$

(b) Again by the chain rule,

$$\frac{d}{dx}[\ln(1-\cos(x))] = \frac{1}{1-\cos(x)}\frac{d}{dx}[1-\cos(x)]$$

$$= \frac{1}{1-\cos(x)}\cdot(0+\sin(x)) = \frac{\sin(x)}{1-\cos(x)}.$$

(c) We could differentiate this directly, using the chain rule (multiple times) and the quotient rule. But it's easier to first simplify the function being differentiated, using properties of logarithms:

$$\ln\left(\frac{x^2}{\ln(x^2)}\right) = \ln(x^2) - \ln(\ln(x^2))$$

$$= 2\ln(x) - \ln(2\ln(x)) = 2\ln(x) - \ln(2) - \ln(\ln(x)).$$

Then

$$\frac{d}{dx}\left[\ln\left(\frac{x^2}{\ln(x^2)}\right)\right] = \frac{d}{dx}[2\ln(x) - \ln(2) - \ln(\ln(x))]$$

$$= \frac{2}{x} - 0 - \frac{1}{\ln(x)}\frac{d}{dx}[\ln(x)] = \frac{2}{x} - \frac{1}{x\ln(x)}.$$

(Since $\ln(2)$ is just a constant, its derivative is zero.)

_____

3.2.4   Exponential Growth and Decay, Revisited

With the logarithm function in hand, we are now able to perform more detailed, complete analyses of exponential growth and decay scenarios. In particular, we can now:

- Solve "how long" problems (like Example 3.2.1(c) above, which we solved in Sect. 3.2.1);
- Evaluate certain parameters, like per capita growth rates and per unit decay rates, given certain additional information about particular values of the quantities in question.
- Evaluate certain related quantities, like "half-lives" (see Example 3.2.4 below) and "doubling times" (see, again, Example 3.2.1(c) and its solution, above).

- Express exponential growth or decay solutions in terms of other bases, which are sometimes more suggestive of the behavior in question.

Here are some examples to illustrate these ideas.

Example 3.2.4 In Exercise 23 of Sect. 1.7.4, we saw that radium 226 has per unit decay rate $k = 1/2337$ year^{-1}. Use this information to find the "half-life," call it τ, of radium 226, meaning the length of time τ that it takes for a given sample to reduce to half of its original mass.

Solution. As we've seen, the amount R of radium 226 present satisfies the differential equation $R' = -(1/2337)R$, if time is measured in years and R is in grams. By what we've seen about exponential decay, then, we know that a sample of R_0 grams of radium reduces to

$$R(t) = R_0 e^{-t/2337} \tag{3.19}$$

grams after t years.

We want to know what time value $t = \tau$ leaves us with half of what we started with. That is, we want to solve the equation $R(\tau) = (1/2)R_0$, or by (3.19),

$$R_0 e^{-\tau/2337} = \frac{1}{2}R_0,$$

for τ. We divide both sides by R_0 to get

$$e^{-\tau/2337} = \frac{1}{2}.$$

Now take natural logarithms of both sides to get

$$-\frac{\tau}{2337} = \ln\left(\frac{1}{2}\right).$$

Finally, solve for τ:

$$\tau = \frac{\ln(1/2)}{-2337} = 1619.88....$$

That is, the half-life of radium 226 is about 1620 years.

Note that, to solve the above example, we **did not** need to know how much radium we started with. We denoted this initial quantity by R_0 but, since we divided through by R_0 along the way, we ended up with a solution that did not depend on R_0. This is characteristic of exponential decay: *if a quantity decays exponentially, then the amount of time it takes for this quantity to halve, or to reduce to one-third of its original amount, or to reduce to p percent of its original amount, for any number p, does not depend on that original* *amount.* (Of course, it does depend on p: you'll be down to 75% of the original amount before you're left with only 10%, and so on.) Similarly for exponential growth: *if a quantity grows exponentially, then the amount of time it takes for this quantity to double, or triple, or grow to a factor of f times the original amount, for any number f, does not depend on that original amount.* (But it does depend on f.)

These claims are true essentially by the arguments used in solving Examples 3.2.1(c) and 3.2.4 above: ultimately, the original amount R_0 (or whatever it's called) gets "divided out" of the equation.

Example 3.2.5 A certain exponentially growing population triples every five years.

(a) What is the per capita growth rate k for this population?
(b) If $P_0 = 10$ (in millions), then (i) find an explicit formula for $P(t)$, with t in years; and (ii) how long does it take for the population to reach 250 million?

Solution. (a) If t is in years and $P(t)$ in millions of people, then

$$P(t) = P_0 e^{kt}, \tag{3.20}$$

where P_0 is also in millions of people.

We're told that $P(5) = 3P_0$ or, by (3.20),

$$P_0 e^{k \cdot 5} = 3P_0.$$

Divide by P_0:

$$e^{5k} = 3.$$

Take the natural logarithm on both sides:

$$5k = \ln(3).$$

Solve for k:

$$k = \frac{\ln(3)}{5} = 0.2197. \tag{3.21}$$

The per capita growth rate is $k = 0.2197 \text{ year}^{-1}$.

(b) (i) By part (a),

$$P(t) = 10e^{0.2197t}$$

millions of people after t years. So: (ii) the population equals 250 million when

$$10e^{0.2197t} = 250.$$

We solve for t by dividing both sides by 10, taking natural logarithms of both sides, and then dividing both sides by 0.2197:

$$e^{0.2197t} = \frac{250}{10} = 25$$

$$0.2197t = \ln(25)$$

$$t = \frac{\ln(25)}{0.2197} = 14.6512.$$

It takes about 14.6512 years for the population to reach 25 million.

The above formula $P(t) = 10e^{0.2197t}$ is useful for calculations. But there is another way of writing this formula, which is more directly reflective of the way $P(t)$ grows, in this case.

Namely: we saw in (3.21) that $k = \ln(3)/5$. So, using properties of exponents and logarithms, we can rewrite the formula $P(t) = 10e^{kt}$ (see (3.20)) as follows:

$$P(t) = 10e^{kt} = 10e^{(\ln(3)/5)t} = 10e^{\ln(3)\cdot(t/5)} = 10(e^{\ln(3)})^{t/5} = 10 \cdot 3^{t/5}. \tag{3.22}$$

This new way of writing $P(t)$ makes it clear that $P(t)$ triples every five years, since the quantity $3^{t/5}$ on the right-hand side of (3.22) becomes three times as large when we increase t by 5. (That is, $3^{(t+5)/5} = 3 \cdot 3^{t/5}$, as you should verify.)

Another thing we should note about the above example is that the result agrees with the following "gut check." We are told that our population triples every five years. So after five years, our original population of 10 million has grown to 30 million. After *another* five years the population of 30 million triples, to 90 million. And after *another* five years, the population of 90 million triples, to 270 million. Thus, after a grand total of $5 + 5 + 5 = 15$ years, our initial 10 million individuals have become 270 million. This is entirely consistent with, and gives us more confidence in, our finding that, after 14.6512 years (a bit less than 15), our population has grown to 250 million (a bit less than 270 million). Or to put it another way: the result $P(14.6512) = 250$ (million) is pretty close to the result $P(15) = 270$ (million), so we feel more comfortable about the former result, even though the calculations there were just a bit messy.

Example 3.2.6 Salt dissolves in water at a rate proportional to the amount $S(t)$ of salt remaining at time t.

If 6 lbs of salt reduce to 5 lbs after one hour, how much salt remains after four hours?

Solution. If we measure t in hours (h) and $S(t)$ in pounds (lb), then we may model our situation by the initial value problem

$$S' = -kS; \qquad S(0) = 6.$$

We know that this problem has solution

$$S(t) = 6e^{-kt} \tag{3.23}$$

for some positive constant k.

We find k by substituting $S(1) = 5$ into (3.23); we get

$$5 = 6e^{-k \cdot 1}.$$

We solve for k:

$$\frac{5}{6} = e^{-k}$$

$$\ln\left(\frac{5}{6}\right) = -k$$

$$k = -\ln\left(\frac{5}{6}\right) = 0.182322 \text{ h}^{-1}. \tag{3.24}$$

So (3.23) reads

$$S(t) = 6e^{-0.182322t}.$$

Finally, we compute:

$$S(4) = 6e^{-0.182322 \cdot 4} = 2.8935 \text{ lb}.$$

Another way of thinking through the above example is this. The equation $k = -\ln(5/6)$, from (3.24), gives us

$$S(t) = 6e^{-kt} = 6e^{-(-\ln(5/6))t} = 6(e^{\ln(5/6)})^t = 6 \cdot (5/6)^t.$$

This latter equation reflects the fact that the amount of salt remaining reduces by a factor of $5/6$ every hour. So after one hour, we have $6 \cdot (5/6) = 5$ lb left; after two hours, we have $6 \cdot (5/6)^2$ lb; after three, we have $6 \cdot (5/6)^3$ lb; after four, we have $6 \cdot (5/6)^4 = 2.8935$ lb, as above.

3.2.5   Exercises

3.2.5.1 Part 1: Basic Properties of the Logarithm Function

1. Determine the numerical value of each of the following. Do not use a calculator for these; just use properties of exponents and logarithms. Your answer to each of the parts of this exercise should be an integer or a rational number (that is, an integer divided by an integer).

(a) $(\ln(e))^5$	(e) $\ln(e^5)$	(i) $4^{\ln(e)}$	(m) $\ln(\sqrt{e})$
(b) $e^{\ln(2)}$	(f) $e^{3\ln(2)}$	(j) $(e^{\ln(2)})^3$	(n) $e^{2\ln(3)}$
(c) $e^{-\ln(2)}$	(g) $e^{-3\ln(2)}$	(k) $5e^{\ln(6/5)}$	(o) $e^{\ln(2)+\ln(3)}$
(d) $e^{\ln(2)+\ln(1/2)}$	(h) $e^{2\ln(2)-3\ln(3)}$	(l) $\ln\left(e^{\ln(e^2)}\right)$	(p) $e^{-\ln(e^{\ln(3)})}$

2. Find dy/dx for each of the following functions.

(a) $y = \ln(3x)$ (d) $y = \ln(2^x)$ (g) $y = x\ln(x)$ (j) $y = x\ln(x) - x$
(b) $y = 17\ln(x)$ (e) $y = \pi\ln(3e^{4s})$ (h) $y = \cos(\ln(x\sin(x)))$ (k) $y = e^{\sqrt{\ln(x)}}$
(c) $y = \ln(e^W)$ (f) $y = \ln(4 + 3x^2)$ (i) $y = 1/\ln(1/x)$ (l) $y = \ln(2)$

3. Use properties of e^x, and ideas like those in the "proof of the second property" in Sect. 3.2.2, to *prove* the following properties of the logarithm. (Remember that $\ln(b) = a$ means $b = e^a$.)

(a) $\ln(1) = 0$.
(b) $\ln(m/n) = \ln(m) - \ln(n)$.
(c) $\ln(m^n) = n\ln(m)$.

3.2.5.2 Part 2: Modeling Growth and Decay with the Exponential Function

4. The rate of growth of the population of a particular country is proportional to the population. The last two censuses determined that the population in 2010 was 308,745,538, and in 2020 it was 331,449,281. What will the population of this country be in 2030? What was the population in 1990?

5. Suppose a bacterial population grows so that its mass is

$$P(t) = 200e^{0.12t} \quad \text{grams}$$

after t hours. Its initial mass is $P(0) = 200\,\text{g}$. When will its mass double, to $400\,\text{g}$? How much longer will it take to double again, to $800\,\text{g}$? After the population reaches $800\,\text{g}$, how long will it take for yet another doubling to happen? What is the *doubling time* of this population?

6. Suppose a beam of X-rays whose intensity is A rads (the "rad" is a unit of radiation) falls perpendicularly on a heavy concrete wall. After the rays have penetrated s feet of the wall, the radiation intensity has fallen to

$$R(s) = Ae^{-0.35s} \quad \text{rads}.$$

What is the radiation intensity 3 in. inside the wall; 18 in.? (Your answers will be expressed in terms of A.) How far into the wall must the rays travel before their intensity is cut in half, to $A/2$? How much further before the intensity is $A/4$?

7. Virtually all living things take up carbon as they grow. This carbon comes in two principal forms: normal, stable carbon—C^{12}—and radioactive carbon—C^{14}. C^{14} decays into C^{12} at a rate proportional to the amount of C^{14} remaining. While the organism is alive, this lost C^{14} is continually replenished. After the organism dies, though, the C^{14} is no longer replaced, so the percentage of C^{14} decreases exponentially over time. It is

found that after 5730 years, half the original C^{14} remains. If an archaeologist finds a bone with only 20% of the original C^{14} present, how old is it?

8. The human population of the world appears to be growing exponentially. If there were 7.21 billion people in 2013, and 7.63 billion in 2018, how many will there be in 2030?

9. If bacteria increase at a rate proportional to the current number, how long will it take 1000 bacteria to increase to 10,000 if it takes them 17 min to increase to 2000?

10. Suppose sugar in water dissolves at a rate proportional to the amount left undissolved. If 40 lb. of sugar reduces to 12 lb. in 4 h, how long will you have to wait until 99% of the sugar is dissolved?

11. Atmospheric pressure is a function of altitude. Assume that at any given altitude the rate of change of pressure with altitude is proportional to the pressure there. If the barometer reads 30 psi (pounds per square inch) at sea level and 24 psi at 6000 ft above sea level, how high are you when the barometer reads 20 psi?

3.2.5.3 Part 3: Differential Equations

12. Find a solution to the initial value problem

$$f'(x) = \frac{3}{x}, \qquad f(1) = 2.$$

Hint: try $f(x) = 3 \ln x + b$, then determine what b needs to be to satisfy the given initial value problem.

13. Find a solution to the initial value problem

$$y' = \frac{a}{x}, \qquad y(1) = b.$$

14. Solve the initial value problem

$$P' = \frac{2}{t}, \qquad P(1) = 5.$$

3.3   Inverse Functions

Much of what we have said about the natural exponential and logarithm functions carries over directly to *any* pair of inverse functions. To explain this, we should first say precisely what it means for two functions f and g to be inverses of each other.

Fig. 3.3 Functions f (in red) and g (in green) that are inverse to each other. The graph of $y = g(x)$ is the graph of $y = f(x)$, reflected about the line $y = x$ (dashed)

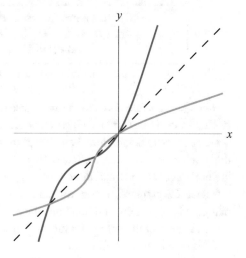

Definition 3.3.1 Two functions f and g are **inverses** if

$$\text{(i) } g(f(a)) = a, \quad and \quad \text{(ii) } f(g(b)) = b, \tag{3.25}$$

for every a in the domain of f and every b in the domain of g.

For example: as previously noted, the functions $f(x) = e^x$ and $g(x) = \ln(x)$ are inverses of each other, by Eqs. (3.10) and (3.16) (Fig. 3.3).

We mentioned in Sect. 3.2 that the functions $f(x) = e^x$ and $g(x) = \ln(x)$ are "flips," or reflections, of each other about the line $y = x$. This is a general property of pairs of inverse functions. That is: suppose we have *any* two functions $f(x)$ and $g(x)$ that are inverses of each other. Of course, the function $f(x)$ takes a number a in its domain to $f(a)$. But by Eq. (3.25)(i), the function $g(x)$ then takes $f(a)$ back to a. Schematically, we have this picture:

$$a \longrightarrow \boxed{\begin{array}{c} \text{the function} \\ y = f(x) \end{array}} \longrightarrow f(a)$$

$$f(a) \longrightarrow \boxed{\begin{array}{c} \text{the function} \\ y = g(x) \end{array}} \longrightarrow a$$

In other words: the function $y = g(x)$ takes the function $y = f(x)$, and **reverses** the roles of input and output.

As noted earlier (in discussing exponential and logarithmic functions), swapping input with output amounts to swapping the horizontal with the vertical. And again, this is the same as *reflecting* everything about the line $y = x$. CONCLUSION:

> **If f and g are inverse functions of each other, then
> the graph of $y = g(x)$ is the graph of $y = f(x)$,
> reflected about the line $y = x$.**

Geometrical relationship between inverse functions

Let $f(x)$ be a function that has an inverse function $g(x)$. We claim that $f(x)$ **must satisfy the horizontal line test**, meaning **no horizontal line can intersect the graph of $f(x)$ more than once.** Why is this true? Well, suppose there *were* a horizontal line $y = b$ intersecting the graph of $f(x)$ more than once. Then, since the graph of $g(x)$ is just that of $f(x)$ with the horizontal and vertical directions swapped, we would find that the *vertical* line $x = b$ intersects the graph of $g(x)$ more than once. But this is impossible, because we've assumed that $g(x)$ is a function, and functions must satisfy the vertical line test. (That is, no vertical line can intersect the graph of a function more than once.)

This motivates the following.

Definition 3.3.2 We say that a function f is **one-to-one**, usually written as **1–1**, if its graph satisfies the *horizontal line test:* no horizontal line intersects the graph more than once.

Here's another way of thinking about 1–1 functions. To say that no horizontal line intersects a graph more than once is to say that no y-value on the graph can come from two different x-values. So: to say that a function $y = f(x)$ is 1–1 is to say that, if inputs x_1 and x_2 are unequal, then the outputs $y_1 = f(x_1)$ and $y_2 = f(x_2)$ must be unequal as well.

We have just seen that only functions that are one-to-one can have inverses. This means that to establish inverses for some functions, we will need to restrict their domains to regions where they are one-to-one. Let's consider some examples.

Example 3.3.1 Suppose $f(x) = x^2$ and $g(x) = \sqrt{x}$. The squaring function $f(x)$ is not invertible on its natural domain because it is not one-to-one there. (It's clear from the graph that $f(x) = x^2$ does not satisfy the horizontal line test. For example, the line $y = 9$ intersects this graph more than once. Or to put it another way, $9 = f(-3) = f(3)$ even though $-3 \neq 3$.)

However, the squaring function $f(x) = x^2$ *is* invertible if we restrict its domain to non-negative real numbers. Then

$$f(g(b)) = \left(\sqrt{b}\right)^2 = b \quad (\text{for } b \geq 0)$$
$$\text{and } g(f(a)) = \sqrt{a^2} = a \quad (\text{for } a \geq 0).$$

The second of these statements—particularly the equation $\sqrt{a^2} = a$—is only true on our restricted domain of f. It fails if we allow a to be negative—for example, $\sqrt{(-3)^2} = \sqrt{9} = 3 \neq -3$.

We already know how to find the derivative of the square root function. But let's compute this derivative again (assuming we know the derivative of the squaring function), to further illustrate how the derivative of an inverse function is related to that of the original function itself.

For our above functions $f(x) = x^2$ and $g(x) = \sqrt{x}$ we have, for appropriate values of x,

$$f(g(x)) = x.$$

Differentiate both sides: the derivative of the right-hand side is just 1, while the derivative of the left-hand side is given by the chain rule. We get

$$f'(g(x))g'(x) = 1$$

or, solving for $g'(x)$,

$$g'(x) = \frac{1}{f'(g(x))}. \tag{3.26}$$

But we can simplify the right-hand side of (3.26): since $f(x) = x^2$, we have $f'(x) = 2x$, so

$$f'(g(x)) = 2g(x) = 2\sqrt{x}.$$

So (3.26) gives $g'(x) = 1/(2\sqrt{x})$ or, recalling that $g(x)$ is the square root function,

$$\frac{d}{dx}[\sqrt{x}] = \frac{1}{2\sqrt{x}},$$

agreeing with earlier results.

Note that we could have restricted the domain of f in another way to make it one-to-one: we could have taken its domain to consist of non-positive real numbers $x \leq 0$, instead of non-negative real numbers $x \geq 0$. Now the function $g(x) = \sqrt{x}$ is no longer the inverse of this restricted f. For instance, $g(f(-3)) = g(9) = 3 \neq -3$. What would the inverse of f be in this case? (See Exercise 3 below.)

Example 3.3.2 Suppose $y = \tan(x)$. Note that this does not define a 1–1 function on the natural domain of the tangent function. For example, $\tan(\pi/4)$ and $\tan(5\pi/4)$ are both equal to 1, even though $\pi/4 \neq 5\pi/4$. (In fact, the the horizontal line $y = 1$ intersects the graph of $y = \tan(x)$ infinitely often, since $\tan(k\pi + \pi/4) = 1$ for every integer k.) So we need to restrict our domain in order for our tangent function to have an inverse.

To this end, let's consider the function $f(x) = \tan(x)$ with domain $-\pi/2 < x < \pi/2$ (Fig. 3.4).

As is clear from the graph, this function *is* 1–1 on the indicated domain. So it has an inverse function $g(x)$ there. This inverse function is called the *arctangent* function, denoted $\arctan(x)$. Since $f(x) = \tan(x)$ has domain $(-\pi/2, \pi/2)$ and range $(-\infty, \infty)$, we find that

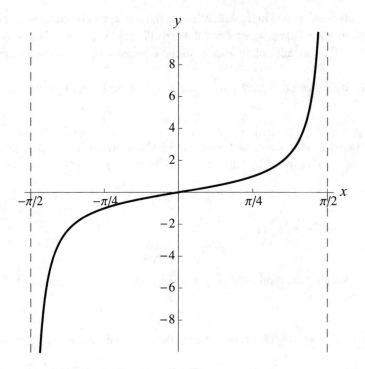

Fig. 3.4 The graph of $f(x) = \tan(x)$ on $(-\pi/2, \pi/2)$

$g(x) = \arctan(x)$ has domain $(-\infty, \infty)$ and range $(-\pi/2, \pi/2)$. The graph of $g(x)$ looks like Fig. 3.5.

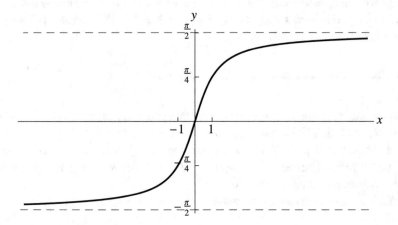

Fig. 3.5 The graph of $g(x) = \arctan(x)$

To find the derivative of the arctangent function, we proceed in the usual way. Specifically: we start with the fact that

$$\tan(\arctan(x)) = x.$$

We differentiate both sides; on the left-hand side, we use the chain rule, and the fact that the derivative of the tangent function is the square of the secant function. We get

$$\sec^2(\arctan(x))\frac{d}{dx}[\arctan(x)] = 1$$

or, dividing both sides by $\sec^2(\arctan(x))$,

$$\frac{d}{dx}[\arctan(x)] = \frac{1}{\sec^2(\arctan(x))}. \tag{3.27}$$

We can simplify the right-hand side using the trigonometric identity

$$\sec^2(\theta) = 1 + \tan^2(\theta).$$

(See Exercise 18, Sect. 1.7.4.) Note that this identity implies

$$\sec^2(\arctan(x)) = 1 + \tan^2(\arctan(x)) = 1 + \big(\tan(\arctan(x))\big)^2 = 1 + x^2, \tag{3.28}$$

the last step because, again, $\tan(\arctan(x)) = x$. Putting (3.28) into (3.27) gives us our ultimate result:

$$\boxed{\frac{d}{dx}[\arctan(x)] = \frac{1}{1 + x^2}}$$

Derivative of the arctangent function

Example 3.3.3 Find (i) $\dfrac{d}{dx}[\arctan(2 + x^5)]$ and (ii) $\dfrac{d}{dr}[2 + \arctan^5(r)]$.

Solution. (i) $\dfrac{d}{dx}[\arctan(2 + x^5)] = \dfrac{1}{1 + (2 + x^5)^2} \cdot \dfrac{d}{dx}[2 + x^5] = \dfrac{5x^4}{1 + (2 + x^5)^2}.$

(ii) $\dfrac{d}{dr}[2 + \arctan^5(r)] = \dfrac{d}{dr}[2 + (\arctan(r))^5] = 0 + 5(\arctan(r))^4 \cdot \dfrac{d}{dr}[\arctan(r)] = \dfrac{5\arctan^4(r)}{1 + r^2}.$

We have seen that, if f and g are inverse functions, then the chain rule gives us a method for finding the derivative $g'(x)$, given knowledge of $f'(x)$. This method was employed

in the context of exponential and logarithmic functions (see Sect. 3.2.3); in the context of
squares and square roots (see Example 3.3.1); and in the context of the tangent and arctangent
functions (see Example 3.3.2). The strategy employed, in each case, was as follows:

Step 1. Begin with the formula $f(g(x)) = x$, where f is the function whose
derivative is already known.

Step 2. Differentiate to get $f'(g(x))g'(x) = 1$.

Step 3. Divide by $g'(x)$ to get $g'(x) = \dfrac{1}{f'(g(x))}$.

Step 4. Simplify the right-hand side if possible. This simplification will often
use the fact that $f(g(x)) = x$.

Strategy for differentiating inverse functions

Step 4 is often the "tricky" step. In the case of exponents and logarithms, or squares and
square roots, this step was fairly straightforward. But in the case of tangents and arctangents,
we needed to make the non-obvious observation that $\sec^2(\theta) = 1 + \tan^2(\theta)$. This allowed
us to obtain an expression involving $\tan(\arctan(x))$, which simplified because the tangent
and arctangent functions are inverse to each other.

Finally: since an inverse to a function $y = f(x)$ (when an inverse exists) is obtained by
interchanging the roles of x and y (that is, of input and output), we can sometimes find an
inverse function, algebraically, by making this interchange—that is, by writing $x = f(y)$—
and solving for y.

Example 3.3.4 To find the inverse of the function

$$y = \frac{3 - x}{2 + x},$$

we interchange x with y:

$$x = \frac{3 - y}{2 + y},$$

and then solve for y:

$$x(2 + y) = 3 - y$$
$$2x + xy = 3 - y$$
$$xy + y = 3 - 2x$$
$$y(x + 1) = 3 - 2x$$
$$y = \frac{3 - 2x}{x + 1}.$$

As a check on our work in the above example, we show that the functions $f(x) = \dfrac{3-x}{2+x}$ and $g(x) = \dfrac{3-2x}{x+1}$ truly are inverses, as follows:

$$
\begin{aligned}
f(g(x)) = f\left(\frac{3-2x}{x+1}\right) &= \frac{3-(3-2x)/(x+1)}{2+(3-2x)/(x+1)} \\
&= \frac{3(x+1)-(3-2x)}{2(x+1)+(3-2x)} = \frac{3x+3-3+2x}{2x+2+3-2x} \\
&= \frac{5x}{5} = x,
\end{aligned}
$$

as required. (For the third "=" in this computation, we multiplied top and bottom by $x + 1$, to simplify.) Strictly speaking, our verification that $f(g(x)) = x$ is valid only for $x \neq -1$, since $g(-1)$ is undefined.

A similar computation shows that $g(f(x)) = x$, for all $x \neq -2$.

Example 3.3.5 Find the inverse g to the function $f(x) = x^2 - 4$ on the domain $x \geq 0$. What is the domain of g?

Solution. We write $y = x^2 - 4$, interchange x and y, and solve for y:

$$
\begin{aligned}
x &= y^2 - 4 \\
y^2 &= x + 4 \\
y &= \pm\sqrt{x+4}
\end{aligned}
$$

We *must* choose the plus sign, because we have stipulated that $f(x)$ have only nonnegative numbers in its domain, so its inverse must have only nonnegative numbers in its range.

So our inverse function is $g(x) = \sqrt{x+4}$. The domain of $g(x)$ is the set of all $x \geq -4$, since this is the range of $f(x)$.

3.3.1   Exercises

3.3.1.1 Part 1: Derivatives Involving the Arctangent Function
1. Find:

(a) $\dfrac{d}{dx}[3\arctan 4x]$

(b) $\dfrac{d}{dx}[e^{\arctan(x)}]$

(c) $\dfrac{d}{dx}\left[\arctan\left(\dfrac{1}{x}\right)\right]$

(d) $\dfrac{d}{dq}[(1+q^2)\arctan(q)-q]$

(e) $\dfrac{d}{dy}[\arctan^3(y\ln(y))]$

(f) $\dfrac{d}{dy}\left[\dfrac{1}{1+\arctan(y)}\right]$

3.3.1.2 Part 2: Finding Inverse Functions
2. Consider the function $f(x)=3x^2-5$ on the domain $x\geq 0$. What is the inverse function to $f(x)$ on this domain? What is the domain of this inverse function?
3. Consider the function $f(x)=x^2$ on the domain $x\leq 0$. What is the inverse function to $f(x)$ on this domain? What is the domain of this inverse function?
4. Show that $f(x)=5-x$ equals its own inverse. What are the domain and range of f?
5. Show that $f(x)=1/x$ equals its own inverse. What are the domain and range of f?
6. Let n be a positive integer. and let $f(x)=x^n$. What is an inverse of f? How do we need to restrict the domain of f for it to have an inverse? Caution: the answer depends on whether n is even or odd.
7. What is the inverse g of the function $f(x)=1-3x$?
8. What is the inverse g of the function $f(x)=\dfrac{1-3x}{2x+5}$?
9. What is the inverse of $f(x)=\dfrac{x^2}{2}+5$ on the domain $x\geq 0$?

3.3.1.3 Part 3: Derivatives of Inverse Functions
10. Use the strategy in the box in Sect. 3.3 and the fact that $d[x^3]/dx=3x^2$ to derive the formula for the derivative of $\sqrt[3]{x}$. (Pretend you don't already know how to differentiate $\sqrt[3]{x}$.) See Example 3.3.1 for a similar problem.
11.
  (a) Use a computer graphing utility to graph $y=\sin(x)$ on the domain $-\pi/2\leq x\leq\pi/2$.
  (b) Explain in words how you can tell from the graph that $y=\sin(x)$ has an inverse function on $[-\pi/2,\pi/2]$.
  (c) Denote the inverse function from part (b) of this exercise by $\arcsin(x)$. What is the domain of $\arcsin(x)$?
  (d) Use Steps 1–3 of the strategy of Sect. 3.3 to show that

$$\frac{d}{dx}[\arcsin(x)]=\frac{1}{\cos(\arcsin(x))}.$$

(See Example 3.3.2 for a similar problem.)

(e) Use the fact that $\cos(x) = \sqrt{1 - \sin^2(x)}$ on the domain $[-\pi/2, \pi/2]$ to conclude that

$$\frac{d}{dx}[\arcsin(x)] = \frac{1}{\sqrt{1 - x^2}}.$$

3.3.1.4 Part 4: Modeling Disease Using the Arctangent Function

The graph below shows a function $D(t)$ given by the formula

$$D(t) = 79\arctan(0.1(t - 45)) + 107.$$

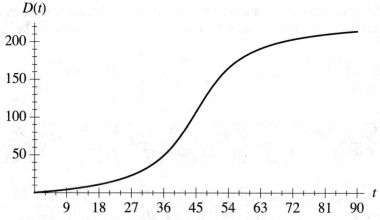

The next graph shows a function $R(t)$ given by the formula

$$R(t) = \frac{7.9}{1 + 0.01(t - 45)^2}.$$

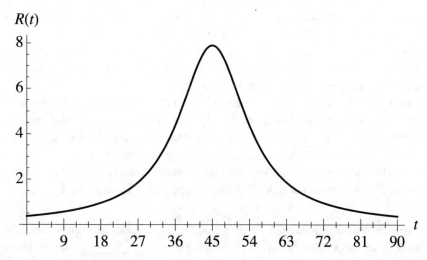

12. Let $D(t)$ and $R(t)$ be the functions given just above. Use the formula

$$\frac{d}{dt}[\arctan(t)] = \frac{1}{1+t^2},$$

together with any relevant differentiation rules and formulas, to show that

$$D'(t) = R(t).$$

We are now going to examine data from an actual 90-day-long outbreak of Ebola in the Democratic Republic of Congo (DRC), in 1995.

Here is a table of deaths due to this outbreak. The top row of the table breaks up the ninety-day period into three-day increments. The bottom row denotes the total, or *cumulative*, number of deaths $C(t)$, due to this Ebola outbreak, from day 0 to the end of day t.

Time t (days)	3	6	9	12	15	18	21	24	27	30
Total deaths $C(t)$	1	1	1	4	11	13	13	17	24	33

Time t (days)	33	36	39	42	45	48	51	54	57	60
Total deaths $C(t)$	33	38	52	88	112	121	135	153	170	182

Time t (days)	63	66	69	72	75	78	81	84	87	90
Total deaths $C(t)$	187	193	195	204	208	210	211	214	214	214

13. Make a copy of the graph of $D(t)$ above. Directly on top of your copy, plot the points given in the above table. That is: plot (with a small circle or dot) each of the points $(t, C(t))$, where $t = 3, 6, 9, \ldots, 90$. (You don't need to connect the dots.)
14. What is one, perhaps surprising, type of phenomenon (seen **very** recently!) that the arctangent function can sometimes be used to model? Please explain your answer, based on what you saw in the previous exercise.
15. Consider the Ebola outbreak documented in the table above, just before Exercise 13. Instead of looking at *cumulative* deaths, let's now consider the *death rate*, measured in deaths per day. What do you think the graph of the death rate function might look like? Please explain. (Hint: it looks like something seen **very** recently!)
16. Fill in the blanks – in each blank *except for the last one*, the correct answer is *either* the word "sigmoid" (which means "roughly S-shaped") or the word "bell."

The cumulative death function for the above Ebola outbreak can be modeled fairly well by an arctangent curve, which has something of a _____ shape. The death

rate for the outbreak can then be modeled fairly well by a curve that has something of
a _____ shape.

So the derivative of a _____ curve is a _____ curve. Where have we
seen this before? We've seen it in SIR!! Remember that, there, the variable R (recovered) followed a _____ curve, while the variable I (infected) followed a
_____ curve. But I is the derivative of R, or more precisely I is *proportional to*
the derivative of R, by the third of the SIR equations, which says $R' =$ _____.

3.4   Modeling Populations

In Sect. 3.1 we saw how population growth, under appropriate conditions, can be modeled
by an exponential function. In this section, we examine this model a bit more closely,
and consider some other models that arise through altering the assumptions behind the
exponential model.

3.4.1   Single-Species Models: Rabbits

The problem. If we turn 2,000 rabbits loose on a large, unpopulated island, how might the
number of rabbits vary over time? If we let $R = R(t)$ be the number of rabbits at time t,
measured in months, say, then we would like to be able to make some predictions about the
function $R(t)$. It would be ideal to have a formula for $R(t)$—but this is not always possible.
Nevertheless, there may still be a great deal we can say about the behavior of R. To begin
our explorations, we will construct a model of the rabbit population that is quite simplistic.
After we analyze the predictions it makes, we'll look at various ways to modify the model
so that it approximates reality more closely.

The first model. Let's assume that, at any time t, the rate at which the rabbit population
changes is simply proportional to the number of rabbits present at that time. We then have
the familiar differential equation

$$R' = kR. \tag{3.29}$$

If t is in months and R in number of rabbits, then the multiplier k, called the **per capita
growth rate** (or the **reproductive rate**), should have units of month^{-1}, so that the units on
the two sides of our differential equation agree with the other.

There is another, perhaps more suggestive, way of understanding k, and its units. We
divide both sides of the differential equation (3.29) by R, to get

$$k = \frac{R'}{R}. \tag{3.30}$$

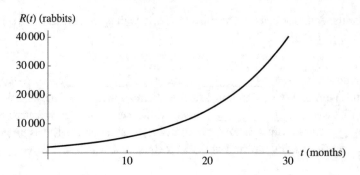

Fig. 3.6 Growth of rabbits according to the initial value problem (3.31)

Since R' has units of rabbits/month, and R has units of rabbits, Eq. (3.30) tells us that k has units of rabbits per month per rabbit. Mathematically, these are the same units as month^{-1}. However, the "rabbits per month per rabbit" interpretation is more indicative of what k measures: it measures how many rabbits per month result from *each* rabbit (on average, of course). Or to put it another way: $1/k$ months is the average length of time required for a rabbit to produce one new rabbit.

For the sake of discussion, let's suppose that $k = 0.1$ rabbits per month per rabbit. We also assume, as above, that there are 2,000 rabbits at the start. We can now state a clearly defined initial value problem for the function $R(t)$:

$$R' = 0.1R, \qquad R(0) = 2,000. \qquad (3.31)$$

As we saw in Sect. 3.1, this initial value problem has solution

$$R(t) = 2,000e^{0.1t} \qquad (3.32)$$

rabbits after t months.

The graph of $R(t)$ looks like this (Fig. 3.6).

This model is too simple to be able to describe what happens to a rabbit population very well. One of the obvious difficulties is that it predicts the rabbit population just keeps growing—forever. For example, if we used the formula for $R(t)$ given above, our model would predict that after 20 years—$t = 240$—there will be $2,000e^{0.1 \cdot 240}$, or more than 50 *trillion,* rabbits! While rabbit populations can, under good conditions, grow at a nearly constant per capita rate for a surprisingly long time (this happened in Australia during the 19th century), our model is ultimately unrealistic.

The second model. One way out of the problem of unlimited growth is to modify Eq. (3.29) to take into account the fact that any given ecological system can support only some finite number of creatures over the long term. This number is called the **carrying capacity** of the system. We expect that, as a population approaches the carrying capacity of the system, the growth of the population should slow way down. That is: near carrying capacity, a population should hold nearly steady—its rate of change should be close to zero.

Let's denote this carrying capacity by b. What we would like to do, then, is to find an expression for R' which is in some ways similar to Eq. (3.29), but which approaches 0 as R approaches b. One model which captures these features is the **logistic equation**, first proposed by the Belgian mathematician Otto Verhulst in 1845:

$$R' = kR\left(1 - \frac{R}{b}\right). \tag{3.33}$$

In this equation, the coefficient k is called the **natural growth rate**. It plays much the same role as the per capita growth rate in Eq. (3.29), and it has the same units (if R and t have the same units as in (3.29)). The carrying capacity b is measured in rabbits (or whatever the units of R are).

Logistic growth—that is, growth according to Eq. (3.33)—has several key features:

1. Suppose that, at some point or interval in time, R is small compared to b. Then the quantity R/b in (3.33) will be small, so that the quantity $1 - R/b$ in parentheses will be close to 1. So (3.33) will "look like" the simpler, exponential growth differential equation $R' = kR$. In other words: as long as the population is much smaller than the carrying capacity, logistic growth looks a lot like exponential growth.
2. Suppose, on the other hand, that R is close to b, but less than b. Then $1 - R/b$ will be close to zero, but positive, so by (3.33), R' will be too. In other words: as the population grows close to the carrying capacity, its growth slows down.
3. If, on the other hand, R is initially *larger* than b, then by (3.33), R' will be negative. So a population that initially exceeds the carrying capacity will decrease—which makes sense because, again, the carrying capacity is the maximum value of R that our system can support. (If R is larger than b, but becomes close to b, then R' will become negative and small, signifying that the attrition in population has slowed.)

Returning to our island of rabbits: for the sake of specificity, let's suppose that the carrying capacity of the island is 25,000 rabbits. If we keep the natural growth rate at 0.1 rabbits per month per rabbit, then our logistic initial value problem for the rabbit population is

$$R' = 0.1R\left(1 - \frac{R}{25,000}\right), \qquad R(0) = 2,000. \tag{3.34}$$

Figure 3.7 depicts the solution to this initial value problem. For comparison, we have also graphed the solution to the exponential growth initial value problem (3.31).

Notice that the two graphs look similar when R is near 2,000, but look quite different later on. (The dashed green line is an "asymptote" for the logistic model—a line that the logistic curve approaches as time elapses.)

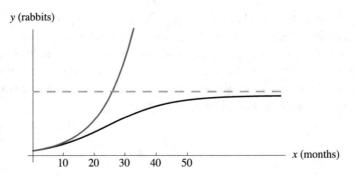

Fig. 3.7 Growth of rabbits according to the logistic (black curve) and exponential (red curve) initial value problems

To graph logistic curves like the one above, one can, of course, apply Euler's method to the logistic Eq. (3.33). Actually though, this equation, like the exponential growth differential equation, admits a closed-form solution. Specifically, one may show that (3.33), together with the initial condition $R(0) = R_0$, have solution

$$R(t) = \frac{R_0 b}{R_0 + (b - R_0)e^{-kt}}. \tag{3.35}$$

We will see how to verify Eq. (3.35) in Exercise 3 below.

3.4.2   Dual-Species Models: Rabbits and Foxes

No species lives alone in an environment, and the same is true of the rabbits on our island. The rabbit population will probably have to deal with predators of various sorts. Some are microscopic—disease organisms, for example—while others loom as obvious threats. We will enrich our population model by adding a second species—foxes—that will prey on the rabbits. Can we say what will happen?

Let $F = F(t)$ denote the number of foxes, and $R = R(t)$ the number of rabbits. As before, we measure the time t in months. We seek differential equations that describe how the growth rates F' and R' are related to the population sizes F and R. We make the following assumptions.

- In the absence of foxes, the rabbit population grows logistically.
- The population of rabbits declines at a rate proportional to the product RF. This is reasonable if we assume rabbits die primarily through predation by foxes. The death rate or rabbits, which depends on the number of fatal encounters between rabbits and foxes, will then be approximately proportional to both R and F—and thus to their product. (This

is the same kind of interaction effect we used in our SIR epidemic model to predict the rate at which susceptibles become infected.)

- Foxes die off at a rate proportional to the number of foxes present.
- Foxes are born at a rate proportional to the number of encounters between rabbits and foxes. To a first approximation, this says that the birth rate in the fox population depends on maternal fox nutrition, and this depends on the number of rabbit-fox encounters, which is proportional to RF.

Our assumptions about birth and death rates can be converted quite naturally into the following differential equations:

$$R' = aR\left(1 - \frac{R}{b}\right) - cRF = aR - \frac{a}{b}R^2 - cRF;$$
$$F' = dRF - eF$$

(Think carefully about how each of our assumptions above corresponds to one of the terms in this pair of differential equations.) These are called the **Lotka–Volterra equations** for bounded (that is, logistic) growth of rabbits. The coefficients a, b, c, d, and e are **parameters**, and are typically determined through field observations in particular circumstances.

Example 3.4.1 To see what kind of predictions the Lotka–Volterra equations make, we'll work through an example with specific values for the parameters. Let

$$
\begin{aligned}
a &= \quad 0.1 \text{ rabbits per month per rabbit} \\
b &= 10,000 \text{ rabbits} \\
c &= \quad 0.005 \text{ rabbits per month per rabbit-fox} \\
d &= 0.00004 \text{ foxes per month per rabbit-fox} \\
e &= \quad 0.04 \text{ foxes per month per fox}
\end{aligned}
$$

(Check that these five parameters have the right units.) Let's also suppose that there are 2,000 rabbits and 10 foxes at time $t = 0$. These choices give us the specific initial value problem

$$
\begin{aligned}
R' &= 0.1\,R - 0.00001\,R^2 - 0.005\,RF, &\quad R(0) &= 2{,}000; \\
F' &= \qquad\quad 0.00004\,RF - 0.04\,F, &\quad F(0) &= 10.
\end{aligned}
\tag{3.36}
$$

Then the two populations will vary in the following way over the next 300 months (Fig. 3.8).

A variant of the program SIR (see Sect. 1.5) was used to produce these graphs. Notice that we have plotted $100F$ rather than F itself. This is because the number of foxes is about 100

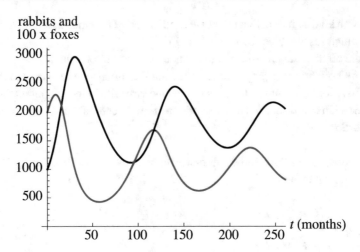

Fig. 3.8 Rabbits (in red) and $100 \times$ foxes (in black), in the two-species model (3.36)

times smaller than the number of rabbits. Consequently, $100F$ and R are about the same size, so their graphs fit nicely together on the same screen.

The graphs have several interesting features. The peak fox population is when $100F$ is about 3,000, meaning $F \approx 30$, while the peak rabbit population is about 2,300. The rabbit and fox populations rise and fall in a regular manner. They rise and fall less with each repeat, though, and if the graphs were continued far enough into the future, we would see R and F level off to nearly constant values.

The illustration below shows what happens to an initial rabbit population of 2,000 in the presence of three different initial fox populations $F(0)$. Note that the peak rabbit populations are different, and they occur at different times. The lengths of the intervals between peaks also depend on $F(0)$ (Fig. 3.9).

We have looked at three models, each a refinement of the preceding one. The first was the simplest. It accounted only for the rabbits, and it assumed the rabbit population grew at a constant per capita rate. The second was also restricted to rabbits, but it assumed logistic growth to take into account the carrying capacity of the environment. The third introduced the complexity of a second species preying on the rabbits. In the exercises, you will have an opportunity to explore these and other models.

3.4.3 Exercises

Most of the exercises at the end of this section, as well as those concluding the next section, will entail the computer implementation of Euler's method, and the outputting of either numerical or graphical information, or both.

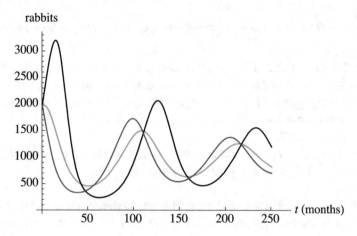

Fig. 3.9 Rabbit populations in the two-species model (3.36), but with initial fox populations $F(0) = 5$ (black), $F(0) = 15$ (green), and $F(0) = 25$ (red)

The code you will need can be obtained by modifying (in most cases, simplifying) the program SIR of Sect. 1.5. It may be helpful to review that section, and especially the exercises at the end of it. Pay particular attention to Exercise 2, Sect. 1.5.2, which describes how to output numerical information (as opposed to graphs).

For some of the Exercises below, you will need to output numerical data, while some call for graphical output. Some entail both kinds of output.

To reach the level of accuracy requested in a given exercise, run your program repeatedly, with successively smaller values of the stepsize Δt, until the decimal digit in question stabilizes (seems no longer to be changing). If accuracy is not specified, choose a stepsize so that smaller stepsizes do not noticeably change the shape of your graph.

3.4.3.1 Part 1: Single-Species Models

1. **Constant per capita growth.** This question considers the initial value problem (3.31):

$$R' = 0.1\,R \text{ rabbits per month}; \qquad R(0) = 2{,}000 \text{ rabbits}.$$

   (a) Use the solution (3.32) to determine how many rabbits there are after 6 months.
   (b) Determine how many rabbits there are after 24 months.
   (c) How many months does it take for the rabbit population to reach 25,000?

2. **Logistic growth, part A: an example.** The following questions concern a rabbit population described by the logistic model

$$R' = 0.1R\left(1 - \frac{R}{25000}\right) \quad \text{rabbits per month;} \quad R(0) = 2,000 \text{ rabbits.}$$

(a) Use Euler's method and Sage to determine, to whole-number accuracy, what happens to this population of 2,000 rabbits after 6 months, after 24 months, and after 6 years.

(b) Modify the program that you used in part (a) (once you've attained the desired level of accuracy) to provide a graph of the solution to the above logistic differential equation, with $R(0) = 2,000$.

(c) Repeat part (b) of this exercise with $R(0) = 40,000$. How does this result differ from that of part (b)? In what ways are the two results similar?

You can check your work against the closed-form solution (3.35) to the logistic equation.

3. **Logistic growth, part B: verification.** Show that, as claimed above, the function

$$R(t) = \frac{R_0 b}{R_0 + (b - R_0)e^{-kt}} \tag{$*$}$$

satisfies the logistic growth initial value problem

$$R' = kR\left(1 - \frac{R}{b}\right), \qquad R(0) = R_0, \tag{$**$}$$

as follows.

(a) Compute $R'(t)$. Hint: by ($*$),

$$R'(t) = \frac{d}{dt}\left[\frac{R_0 b}{R_0 + (b - R_0)e^{-kt}}\right] = R_0 b \cdot \frac{d}{dt}\left[\frac{1}{R_0 + (b - R_0)e^{-kt}}\right],$$

since $R_0 b$ is constant with respect to t. Compute the derivative on the right-hand side using the reciprocal rule.

(b) Compute, and simplify,

$$kR(t)\left(1 - \frac{R(t)}{b}\right).$$

Hint: obtain a common denominator.

(c) Show that your answers to parts (a) and (b) of this exercise are equal, thereby verifying that the function $R(t)$ satisfies the differential equation in ($**$).

(d) Show that $R(t)$ satisfies the initial condition in ($**$) by plugging $t = 0$ into ($*$).

(e) As an additional step (not part of the initial value problem ($**$) *per se*), use ($*$) to determine what happens to $R(t)$ as time evolves indefinitely—that is, as $t \to +\infty$. (Hint: as $t \to +\infty$, e^{-kt} approaches zero.) Your answer should involve a single parameter.

Explain how your result agrees with what you already knew about carrying capacity.

4. **Seasonal Factors.** Living conditions for most wild populations are not constant through-out the year—due to factors like drought or cold, the environment is less supportive during some parts of the year than at others. Partially in response to this, most animals don't reproduce uniformly throughout the year. This problem explores ways of modifying the logistic model to reflect these facts.

For the eastern cottontail rabbit, for example, most young are born during the months of March–May, with reduced reproduction during June–August, and virtually no repro-duction during the other six months of the year. In such a situation, we might have an initial value problem that looks like this:

$$R' = kR\left(1 - \frac{R}{25,000}\right), \qquad R(0) = 2,000,$$

where

$$k = \begin{cases} 0 & \text{during January and February;} \\ 0.2 & \text{during March, April, and May;} \\ 0.05 & \text{during June, July, and August;} \\ 0 & \text{the rest of the year.} \end{cases}$$

(a) Modify your logistic population growth program from Exercise 2(b) above, to account for the varying-growth-rate situation described here. Run this program from $t = 0$ through $t = 24$ months (two years). Hint: add the following code in the appropriate place, appropriately indented:

```
if (t>=0 and t<2) or (t>=12 and t<14):
    k=0
elif (t>=2 and t<5) or (t>=14 and t<17):
    k=0.2
elif (t>=5 and t<8) or (t>=17 and t<20):
    k=0.05
elif (t>=8 and t<12) or (t>=20 and t<24):
    k=0
```

(The Sage term elif is short for "else, if," meaning "otherwise, if.") Reflect on how these lines of code are working, so that you may modify them for the following parts of this exercise.

(b) Modify the code from part (a) of this exercise to extend out through four years ($t = 48$).

(c) How would you modify your code from part (b) to take into account the fact that rabbits don't reproduce during their first season?

5. **World population.** The world's population in 1990 was about 5 billion, and data show that birth rates range from 35 to 40 people per year per thousand people and death rates

from 15 to 20 people per year per thousand people. Take this to imply a net annual growth rate of 20 people per year per thousand people. One model for world population assumes constant per capita growth, with a per capita growth rate of $20/1000 = 0.02$.

(a) Write a differential equation for world population P, measured in billions, that expresses this assumption.
(b) According to the differential equation in (a), at what rate (in billions of persons per year) was the world population growing in 1990?
(c) Using the initial value of 5 billion in 1990, estimate the world population in the years 1980, 2000, 2040, and 2230.

3.4.3.2 Part 2: The Lotka–Volterra Equations with Exponential Growth

This model for predator and prey interactions is slightly simpler than the "bounded growth" version we consider in the text. It is important historically, though, because it was one of the first mathematical population models, proposed as a way of understanding why the harvests of certain species of fish in the Adriatic Sea exhibited cyclical behavior over the years. For the sake of variety, let's take the prey to be hares and the predators to be lynx.

Let $H(t)$ denote the number of hares at time t and $L(t)$ the number of lynx. This model, the basic Lotka–Volterra model, differs from the bounded growth model in only one respect: it assumes the hares would experience constant per capita growth if there were no lynx.

6. (a) Explain why the following system of equations incorporates the assumptions of the basic model. (The parameters a, b, c, and d are all positive.)

$$H' = a\,H - b\,HL$$
$$L' = c\,HL - d\,L$$

(These are called the **Lotka–Volterra equations**. They were developed independently by the Italian mathematical physicist Vito Volterra in 1925–26, and by the mathematical ecologist and demographer Alfred James Lotka a few years earlier. Though simplistic, they form one of the principal starting points in ecological modeling.)

(b) Explain why a and b have the units hares per month per hare and hares per month per hare-lynx, respectively. What are the units of c and d? Explain why.

Suppose time t is measured in months, and suppose the parameters have values

$$a = 0.1, \quad b = 0.005, \quad c = 0.00004, \quad d = 0.04.$$

This leads to the system of differential equations

$$H' = 0.1H - 0.005HL$$
$$L' = 0.00004HL - 0.04L.$$

(c) Suppose that you start with 2,000 hares and 10 lynx—that is, $H(0) = 2,000$ and $L(0)$ = 10. Describe what happens to the two populations. A way to get a good scale on this is to draw graphs of the functions $H(t)$ and $60L(t)$ on the same set of axes. (Use a stepsize of 0.01.)

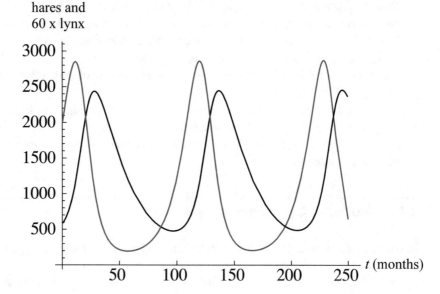

hares and
60 x lynx

You should get a graph like the one above. (Here, hares are in red and 60 × lynx in black.) Notice that the hare and lynx populations rise and fall in a fashion similar to the rabbits and foxes of Example 3.4.1 above, but here they oscillate—returning *periodically* to their original values.

(d) What happens if you keep the same initial hare population of 2,000, but use different initial lynx populations? Try $L(0) = 20$ and $L(0) = 50$. (In each case, use a stepsize of 0.01 month.)

(e) Start with 2,000 hares and 10 lynx. From part (c), you know the solutions are periodic. The goal of this part is to analyze this periodic behavior. First find the maximum number of hares. What is the length of one **period** for the hare population? That is, how long does it take the hare population to complete one cycle (e.g., to go from one maximum to the next)? Find the length of one period for the lynx. Do the hare and lynx populations have the same periods?

(f) Plot the hare populations over time when you start with 2,000 hares and, successively, 10, 20, and 50 lynx. Is the hare population periodic in each case? What is the period? Does it vary with the size of the initial lynx population?

7. Let all quantities be as in the previous exercise. Instead of studying L as a function of t, or H as a function of t, let's now consider L as a function of H!

What do you think you'd see? Try carefully to picture what a graph of L versus H might look like. Then create such a graph. At this point, doing so is relatively straightforward: you can replace any graphics commands you used in part (c) of the previous exercise (that is, any commands following your loop) with a single line, like this:

```
list_plot(list(zip(Hvalues, Lvalues)), axes_labels=['$H$', '$60L$'])
```

(Here, `Hvalues` would be the list in which you have stored your values of H, and `Lvalues` your values of $60L$.)

8. What would you see if you were to take the foxes F and rabbits R of Example 3.4.1, and plot F against R, or $100F$ against R (in a fashion similar to that of the Exercise 7 directly above)? Write a program to do this, and comment on how well your predictions hold up. (Or what surprises you!)

3.5   Modeling Other Phenomena

In this section, we consider various other systems that can be modeled using initial value problems.

Throughout, our emphasis will be on how assumptions about the behavior of a system can be translated into rate equations. And we will take an analytic, "term-by-term" approach. In other words we will endeavor, where possible, to demonstrate how each assumption corresponds to one or more terms in one or more of the rate equations involved.

We will certainly also give some justification for, or explanation of, each of the assumptions. Still, our focus will be less on debating the validity of the assumptions themselves than it will be on accepting a given set of assumptions and deriving a mathematical model from it. This is not to say that careful analysis of the assumptions is not important. It *is* important—it's essential—in science. But it's not our focus.

3.5.1   Circadian Rhythms

"Circadian" means "occurring naturally on a 24-hour cycle."

In 2017, US scientists Jeffrey Hall, Michael Rosbash, and Michael Young received the Nobel Prize in Physiology or Medicine, for their discoveries concerning circadian rhythms. Central to their work was analysis of two genes—a "period" gene and a "timeless" gene— and of the proteins expressed by these genes. (The work of Michael Young also concerned a third gene, which he named the "doubletime" gene. For simplicity, though, we will study only the first two genes.)

Here we consider some earlier studies that laid some of the groundwork for the Nobel-Prize-winning research. In particular, we reference the article "A Simple Model of Circadian Rhythms Based on Dimerization and Proteolysis of PER and TIM," by John J. Tyson, Christian I. Hong, C. Dennis Thron, and Bela Novak, in *Biophysical Journal*, Volume 77, November 1999, pages 241–2417. We focus especially on the differential equations developed in this article to model circadian rhythms.

According to this article, "body clocks" (in fruit flies) are regulated by the *feedback* of two proteins, PER—short for "periodic"—and TIM—short for "timeless"—on the "per" and "tim" mRNA (messenger RNA) that express these proteins.

This feedback loop may be expressed in terms of three basic variables:

- the concentration of per and tim mRNA (taken together), denoted by M;
- the concentration P_1 of PER and TIM *monomers*. A monomer is a basic building block of proteins.
- the concentration P_2 of PER and TIM *dimers*. A dimer is two monomers joined together.

Remark 3.5.1 Here we have followed the authors' practice in grouping together the per and tim mRNA, as well as the PER and TIM monomers and the PER and TIM dimers. As the authors note, one could alternatively separate per from tim, and PER from TIM, resulting in a system of six separate variables and a corresponding system of six differential equations. But the authors argue that "Such a complicated set of equations would not effectively illustrate the importance of positive feedback in the reaction mechanism." In other words, many salient results can be obtained by considering only the variables M, P_1, and P_2 described above.

The authors' model results in a system of differential equations that we will first present, and then explain in terms of the assumptions behind the model. The system, which we denote by (CR) (for "Circadian Rhythm"), is:

$$\frac{dM}{dt} = \underset{\text{(i)}}{-cM} + \underset{\text{(ii)}}{\frac{a}{1 + bP_2^2}}$$

$$\frac{dP_1}{dt} = \underset{\text{(iii)}}{dM} - \underset{\text{(iv)}}{\frac{eP_1}{f + P_1 + gP_2}} - \underset{\text{(v)}}{hP_1} \underset{\text{(vi)}}{- 2kP_1^2} + \underset{\text{(vii)}}{2\ell P_2} \qquad \text{(CR)}$$

$$\frac{dP_2}{dt} = \underset{\text{(viii)}}{- \frac{mP_2}{f + P_1 + gP_2}} - \underset{\text{(ix)}}{nP_2} + \underset{\text{(x)}}{kP_1^2} - \underset{\text{(xi)}}{\ell P_2}$$

The Circadian Rhythm (CR) system of differential equations

Here, the letters $a, b, c, d, e, f, g, j, k, \ell$ all indicate positive parameters. Also, the units of t are hours, while the units of P_1, P_2, and M are certain "characteristic concentrations" for mRNA and protein, respectively.

Notice that we have marked each term, in the above system of equations, with a lower-case roman numeral. (If a term has a minus sign in front of it, that minus sign is considered part of the term.) We now explain the implications of each of these terms.

(i) This term indicates that mRNA is degraded, or used up, at a rate proportional to the amount of mRNA present.

(ii) This term indicates the fact that P_2 *inhibits* growth of M. Indeed, as P_2 increases, the denominator $1 + bP_2^2$ increases, so that the term $a/(1 + bP_2^2)$ decreases. And since this term contributes to dM/dt, the ultimate effect of an increase in P_2 is a decrease in dM/dt, and therefore a slowing down, or inhibition, of the growth of M.

Note that the term (ii) involves the square of P_2, rather than P_2 to the first power. The exponent 2 here is called the "Hill coefficient," and measures certain biochemical binding properties of the molecules involved.

A larger Hill coefficient indicates a greater rate of inhibition. This is because, the larger the exponent h, the faster P_2^h grows as a function of P_2. So a larger h means a smaller $a/(1 + bP_2^h)$, and therefore, in this case, a smaller dM/dt.

We also note that the "1" in the denominator of term (ii) prevents the possibility of division by zero: if the denominator of (ii) were simply bP_2^2, or P_2^2, then this denominator would be zero whenever P_2 is zero. But as long as P_2 is nonnegative (and a negative concentration does not make sense), then $1 + bP_2^2 > 0$.

(iii) This term indicates that PER/TIM monomers are expressed by per/tim mRNA at a rate proportional to the amount of mRNA present.

(iv) and (viii) These are the most complicated terms. They represent a phenomenon known as *phosphorylation*. This is a process whereby both monomers and dimers combine with phosphates and, through this combination, are deactivated.

To understand what these terms are indicating, let's consider their numerators and denominators separately. Were the denominator in term (iv) not present—that is, were term (iv) simply to equal $-eP_1$—then this term would be telling us that monomers are deactivated at a rate proportional to the amount of monomers present. Similarly, without the denominator, term (viii) would imply that the rate of dimer deactivation is proportional to the amount of dimers present. Such behavior on its own would make sense in much the same way that radioactive decay makes sense: if a substance decays, degrades, or deactivates at a certain rate per unit of that substance, then the more of that substance is present, the more of that substance will cease to exist in a given amount of time.

However, the situation here is more complicated than that of radioactive decay: it turns out that, at the same time, protein monomers and dimers inhibit their own decay. This inhibitive effect is captured by the denominators $f + P_1 + gP_2$ in terms (iv) and (viii). (Again, these denominators get larger as P_1 or P_2 does.)

(v) and (ix) These terms represent a process known as *proteolysis*, which is a simpler type of degradation of proteins.

(vi) and (x) These terms represent "dimerization," or the combination of two monomers to form a dimer.

Observe that term (vi) has coefficient $-2k$, whereas (x) has coefficient $+k$. This can be explained as follows:

(a) The coefficient of (vi) is negative, while that of (x) is positive, because dimerization means we are *losing* monomers and *gaining* dimers.

(b) Term (vi) involves a "$2k$," while term (x) involves only a single "k," because for every dimer gained, two monomers are lost.

Also observe that, in (vi) and (x), the variable P_1 occurs to the *second* power. This is because a dimer results from the combination of a monomer with another monomer, so the rate at which dimerization occurs is proportional to the number of *possible* monomer-to-monomer interactions. But the number of such interactions is itself proportional to M^2. (Think of it this way: If there are M people in a room, then there are roughly M^2 possible handshakes that can happen, since every one of the M people can shake hands with any of one M others. Actually this is not quite true, because we should exclude handshakes with oneself—just as a monomer can't combine with itself. But there are relatively few such self-interactions compared to M^2, so that accounting for them or not will make a relatively small difference.)

(vii) and (xi) represent the process where a dimer splits into monomers. Through this process, each dimer lost results in two dimers gained, and this explains the "$+2\ell$" in term (vii) versus the "$-\ell$" in term (xi).

Note that, in (vii) and (xi), the variable P_2 appears only to the *first* power: a dimer does not have to interact with another to split into monomers, so the number of *possible* dimer splits is only proportional to P_2, not to P_2^2. Contrast this with the analysis of terms (vi) and (x) above.

One nice feature of the above model is that it allows us to see quite clearly how certain natural aberrations can affect circadian rhythms.

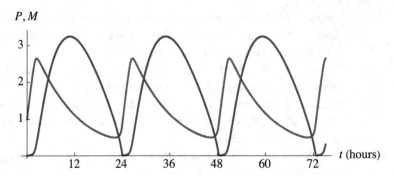

Fig. 3.10 Circadian rhythms under "normal" conditions: mRNA (in red) and protein (in blue)

To illustrate this most simply, it will be helpful first to combine monomers and dimers into a single variable P (for "proteins"). As the authors of the paper show, we can, under certain "equilibrium" conditions, then also combine the above differential equations defining dP_1/dt and dP_2/dt into a single equation for dP/dt. We skip the mathematical details here—they are not completely beyond the scope of this book, but are a bit messy. (These details may be found in the original article.)

The result of this combination is a system of *two* differential equations, one for mRNA M and one for proteins P. Specifying known "normal" values for the parameters in question, we may then create plots of P and M together, using a variant of our program SIR. The resulting graph looks like Fig. 3.10.

Note that the phenomena represented by Fig. 3.10 are circadian: the graphs of both M and P repeat themselves every 24 h. (Actually, the period of these graphs turns out to be closer to 24.2 h, reflective of the fact that, in nature, circadian rhythms are often not precisely circadian, even under "normal" conditions.)

But now we consider an anomalous situation. It's known that a certain *mutation*, denoted per^L, of per mRNA causes a change in the "dimerization rate." By the latter, we mean the parameter k that appears in terms (vi) and (x) above, and that governs the rate at which monomers combine to form dimers. Graphing M and P with a value of k that corresponds to this mutation then yields a result like Fig. 3.11.

The phenomena represented by this graph are quite far from circadian!

We have endeavored to understand the system (CR) by studying each term in the right-hand sides of the differential equations there. In our analysis of these terms, we have encountered a variety of important notions, some of which we've seen previously:

- Notions of growth and decay, generally indicated by "plus" and "minus" signs, respectively.

  The convention that all parameters be positive, as imposed above, helps us insure that plus and minus signs do, in fact, indicate growth and decay, respectively. For example: since

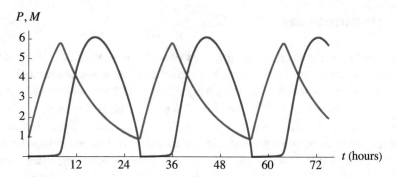

Fig. 3.11 Circadian rhythms under "mutant" conditions: mRNA (in red) and protein (in blue)

M must always be nonnegative, requiring that c be positive means that we immediately recognize the term $-cM$ (cf. term (i) above) as representative of decay, rather than growth. We encountered this convention earlier, in our discussions of exponential growth and decay in Sect. 3.1. There, we stipulated that the parameter k, which represented a growth rate or a decay rate depending on the situation, always be positive. Such a requirement allows us to immediately recognize the equation $P' = kP$ as representative of a growth situation, and the equation $P' = -kP$ as representative of a decay situation (since the quantity P in question generally does not assume negative values).

- The notion of proportionality, where a term in the differential equation for one quantity equals a constant times that quantity, or times some other quantity;
- The notion of inhibition of one quantity on the growth of another, generally reflected by the former quantity appearing in the *denominator* of a term in the differential equation for the latter;
- The notion of exchange, where the transformation of certain quantities into others is reflected by "matching" terms in the differential equations for the quantities involved. (Those terms don't necessarily cancel *per se*. For example: again, since one dimer can become two monomers, the above term (vii) cancels against *twice* the term (xi).)
- The notion of *combinatorics*, where we *count* the number of interactions of a certain type, and this count determines the exponents of the variables in the corresponding terms of the differential equations. (See the analysis of terms (vi) and (x), and terms (vii) and (xi), above.)

Many of these notions arise frequently in the modeling of phenomena by differential equations.

3.5.2 Neural Impulses

There are many mathematical models for *action potential*, meaning a rapid rise and fall—a "spike"—in voltage across a cell membrane. Here we consider the seminal model developed by Sir Alan Hodgkin and Sir Andrew Huxley in the 1940's. This model was the first to exhibit truly predictive power, and laid the groundwork for a vast amount of subsequent research on neural impulses. For their work, Hodgkin and Huxley received the 1963 Nobel Prize in Physiology and Medicine (shared with Sir John Eccles, for his work on transmission across a synapse). Throughout this subsection, we will use lowercase letters to represent parameters (always assumed positive), as well as the independent variable t; dependent variables will be denoted by upper-case letters.

We begin our analysis by considering several fundamental quantities: charge Q, capacitance c, which is a measure of charge *storage capacity* across a cell membrane, and voltage V, which measures the amount of *work* required to move a charge against an electric field. These three quantities are related by the basic equation

$$Q = cV. \tag{3.37}$$

Next, *current I* is defined as the rate of change of charge, with respect to time: $I = \dfrac{dQ}{dt}$ or, by Eq. (3.37),

$$I = \frac{d}{dt}[cV] = c\frac{dV}{dt}. \tag{3.38}$$

In the Hodgkin-Huxley model, I comprises four components:

$$I = I_E - I_{Na} - I_K - I_L, \tag{3.39}$$

where:

- I_E represents an externally applied current;
- I_{Na} represents sodium ion current flowing through sodium channels in the cell membrane;
- I_K represents potassium ion current flowing through potassium channels in the cell membrane;
- I_L represents "leakage" current (coming mostly from chloride ions), due to natural permeability of the cell membrane.

So we can rewrite Eq. (3.38):

$$c\frac{dV}{dt} = I_E - I_{Na} - I_K - I_L. \tag{3.40}$$

Let's now look more closely at each of the four terms on the right-hand side of Eq. (3.40).

(A): I_E. The externally applied current is assumed to be known (and is often constant or zero).

(B): I_{Na}. By *Ohm's Law*, we know that

$$I_{Na} = G_{Na}(V - e_{Na}). \tag{3.41}$$

Here, G_{Na} is *conductance* of the sodium channels (a measure of "compliance" of these channels to current flow), and e_{Na} is the "equilibrium potential" for these channels. (By Eq. (3.41), e_{Na} is the voltage that makes I_{Na} equal to zero.)

To better understand Eq. (3.41), let's study G_{Na} more closely.

Each particular sodium channel comprises four so-called "gates," which are possible avenues for the passage of sodium ions. These gates are of two types. Three of these gates are known as "activation gates," and all three activation gates have the same probability, call it M, of being open, and allowing sodium ions through. The fourth gate is an "inactivation gate," and this fourth gate has a different probability, call it H, of letting sodium ions through. An important point to make here is that *M and H are variables; they change with time.*

Because of the way probabilities multiply, this means that the probability of sodium ions flowing through all four gates of any given sodium channel is $M^3 H$. But there are many sodium channels. Let g_{Na} denote the *maximum possible conductance*, meaning conductance when all gates in all sodium channels are open. We can conclude that the actual conductuance G_{Na} then satisfies the equation

$$G_{Na} = g_{Na} M^3 H. \tag{3.42}$$

Combining Eqs. (3.41) and (3.42) yields the equation

$$I_{Na} = g_{Na} M^3 H (V - e_{Na}). \tag{3.43}$$

(C): I_K. The analysis of I_K is similar to that of I_{Na}. But it's simpler, because all four potassium ion gates are of the same type. Let's denote each gate's "permissivity probability," or probability of being open, by N—which, like M and H above, is a function of time. Then by arguments like those in part (B) above,

$$I_K = g_K N^4 (V - e_K), \tag{3.44}$$

where e_K is the equilibrium potential for the potassium channels, and g_K is the maximum possible conductance of these channels, meaning the conductance that would result were all gates in all potassium channels open.

(D): I_L. The analysis here is even simpler, because the maximum possible conductance due to leakage turns out to be a constant, call it g_L. So

$$I_L = g_L (V - e_L), \tag{3.45}$$

where e_L is the equilibrium potential for leakage current.

We now put Eqs. (3.43), (3.44), and (3.45) into Eq. (3.40), to get

$$c\frac{dV}{dt} = I_E - g_{Na}M^3 H(V - e_{Na}) - g_K N^4(V - e_K) - g_L(V - e_L). \tag{3.46}$$

The quantities M, H, V, and N are the dependent variables here. And (3.46) is, of course, an equation for dV/dt. To complete our model, then, we'd like to develop equations for dM/dt, dH/dt, and dN/dt.

We first consider H, the probability of a sodium inactivation gate being permissive (allowing the flow of sodium ions). Such a gate has probability $1 - H$ of being impermissive. Now suppose we know that such a gate transitions from the permissive state to the impermissive state at the rate A_H, and transitions in the reverse direction (from impermissivity to permissivity) at the rate B_H. We can conclude that

$$\frac{dH}{dt} = A_H H + B_H(1 - H). \tag{3.47}$$

Similarly,

$$\frac{dM}{dt} = A_M M + B_M(1 - M) \tag{3.48}$$

and

$$\frac{dN}{dt} = A_N H + B_H(1 - H). \tag{3.49}$$

Equations (3.46), (3.47), (3.48), and (3.49) *are* the differential equations of the Hodgkin-Huxley model. For clarity, we present them all together, here:

$$c\frac{dV}{dt} = I_E - g_{Na}M^3 H(V - e_{Na}) - g_K N^4(V - e_K) - g_L(V - e_L)$$

$$\frac{dH}{dt} = A_H H + B_H(1 - H)$$

$$\frac{dM}{dt} = A_M M + B_M(1 - M) \tag{HH}$$

$$\frac{dN}{dt} = A_N N + B_N(1 - N)$$

The Hodgkin-Huxley (HH) system of differential equations

The above system entails a number of parameters: g_{Na}, e_{Na}, g_K, e_K, g_L, and e_L. But note also that there are a number of *dependent variables* appearing on the right-hand sides of these equations, beyond the ones appearing on the left-hand sides. Specifically, there are the variables I_E, A_H, B_H, A_M, B_M, A_N, and B_N. To make equations (HH) into a viable system—one that we could, for example, solve using Euler's method (together with a set of given initial conditions), we would need to know more about these variables. It's already been noted that I_E will, generally (at least in our present model), be known or specified. Further, relatively simple formulas for the functions A_H, B_H, A_M, B_M, A_N, and B_N, *in terms of the voltage V*, can also be given, based on experimentation and various assumptions. See, for example, the chapter "Electrophysiological Models," by M. E. Nelson, in the text *Databasing the Brain: From Data to Knowledge* (S. Koslow and S. Subramaniam, eds.), Wiley, New York (2004).

3.5.3 Exercises

3.5.3.1 Part 1: Fermentation

Wine is made by yeast, in the following way. Yeast digests the sugars in grape juice, and produces alcohol as a waste product. This process is called fermentation. The alcohol is toxic to the yeast, though, and the yeast is eventually killed by the alcohol. This stops fermentation, at which point the liquid has become wine, with about 8–12 percent alcohol.

The following exercises develop a sequence of models to take into account the interactions between sugar, yeast, and alcohol.

1. In this exercise, you will build a differential equation for yeast, and plot the solution to this differential equation.

    (a) In this first model, assume that **yeast grows logistically**, with carrying capacity equal to 10 lbs of yeast. Assume that the natural growth rate of the yeast is 0.2 lbs of yeast per hour, per pound of yeast. Let $Y(t)$ be the number of pounds of live yeast present after t hours; what differential equation describes the growth of Y? (You may want to review the discussion of logistic growth in Sect. 3.4 above.)
    (b) Using Euler's method and Sage, sketch a graph of the solution $Y(t)$ to the differential equation from part (a) above, subject to the initial condition $Y(0) = 0.5$ lb of yeast.

2. In this next exercise, you will consider the effects of alcohol and yeast on each other.

    (a) Now consider how the yeast produces alcohol. Suppose that **alcohol is generated at a rate proportional to the amount of yeast present**. Specifically, suppose each pound of yeast produces 0.05 lbs of alcohol per hour. (The other major waste product is carbon dioxide gas, which bubbles out of the liquid, and will not be considered further here.) Let $A(t)$ denote the amount of alcohol generated after t hours. Construct a differential equation that describes the growth of A.
    (b) Now consider the toxic effect of the alcohol on the yeast. Assume that **yeast cells die at a rate proportional to the amount of alcohol present, times the amount of yeast present**. Specifically, assume that, in each pound of yeast, a pound of alcohol will kill 0.1 lb of yeast per hour. Then, if there are Y lbs of yeast and A lbs of alcohol, how

many pounds of yeast will die in one hour? Modify the original logistic equation for
Y (from Exercise 1 above) to take this effect into account. The modification involves
subtracting off a new term that describes the rate at which alcohol kills yeast. What
is the new equation for Y'?

(c) You should now have two differential equations, describing the rates of growth of
yeast and alcohol. The equations are coupled, in the sense that the yeast equation
involves alcohol, and the alcohol equation involves yeast.

Assume that the vat contains, initially, 0.5 lb of yeast and no alcohol. Modify your
code from Exercise 1, above, to generate a graph of both Y and A, on the same set of
axes, as functions of t. On your graph, make sure you indicate clearly which curve
is Y and which is A. (For example, you can do this using `legend_label` and
`legend_color`. See the last few lines of the SIR program from Sect. 1.5.1.)

3. Finally, in this exercise, you will consider the effect of sugar in the system. To complete
part (c) of this exercise, you will need to modify your code from Exercise 2 above.

(a) The third model will take into account that the sugar in the grape juice is consumed by
the yeast. Suppose that **sugar is consumed at a rate proportional to the amount of
yeast present**. (Note that this is telling us something about S', not Y'.) Specifically,
assume that the yeast consumes 0.15 lb of sugar per hour, per lb of yeast. Let $S(t)$
be the amount of sugar in the vat after t hours. Write a differential equation that
describes what happens to S over time.

(b) We will now account for the fact that the carrying capacity for yeast actually *depends*
on the amount of sugar present (so that this carrying capacity now varies with time).
Let's assume that **the carrying capacity for yeast is proportional to the amount
of sugar present**. Specifically, we assume that the carrying capacity for yeast equals
$0.4S$ lbs, where S is the amount of sugar present at that time. Rewrite the logistic
equation for Y so that the carrying capacity for Y is $0.4S$ lbs, instead of 10 lbs. In
this new equation, you should retain the term, developed in Exercise 2(b) above, that
reflects the toxic impact of alcohol on the yeast. What is your new equation for Y'?

(c) There are now three differential equations. Using Sage, produce a graph that depicts
what happens to Y, A, and S over time, subject to these differential equations. Assume
starting values $Y(0) = 0.5$, $S(0) = 25$, and $A(0) = 0$ (all in lb).
Warning: Be careful when typing in your equation of the form `Yprime=....` If you
simply input `Y/b*S` into Sage, it will be interpreted as `(Y/b)*S`, which is not what
you want. So make sure you use parentheses and write `Y/(b*S)`, which is what part
of your Sage equation for `Yprime` should look like.
In your output, make sure you indicate clearly which curve is Y, which is A, and
which is S.

3.5.3.2 Part 2: Newton's Law of Cooling

4. In our discussion of Newton's Law of Cooling—see Exercise 24, Sect. 1.7.1, and Exercise 12, Sect. 3.1.5—we assumed that the coffee did not heat up the room. This is reasonable because the room is large, compared to the cup of coffee. Suppose, in an effort to keep it warmer, we put the coffee into a small insulated container—such as a microwave oven (which is turned off). We must assume that the coffee *does* heat up the air inside the container. Let A be the air temperature in the container and C the temperature of the coffee. Then both A and C change over time, and Newton's law of cooling tells us the *rates* at which they change. In fact, the law says that both C' and A' are proportional to $C - A$. Thus,

$$C' = -k_1(C - A)$$
$$A' = k_2(C - A),$$

where k_1 and k_2 are positive constants.

(a) Explain the signs that appear in these differential equations.
(b) Suppose $k_1 = 0.3$ and $k_2 = 0.1$. If $C(0) = 90\,°C$ and $A(0) = 20\,°C$, when will the temperature of the coffee be $40\,°C$? What is the temperature of the air at this time? Use Euler's method to answer. Your answers should be accurate to one decimal place.
(c) What does the temperature of the coffee become eventually? How long does it take to reach that temperature?

3.5.3.3 Part 3: A Genetic Toggle Switch

In the following exercises, we develop a a dynamical system to model a *bistable genetic toggle switch*. We explain: by "genetic toggle switch," we mean a genetic system that can be used to switch on or off production of a specific protein. By "bistable," we mean the switch stays on once switched on, and stays off once switched off (even if the "inducer," or stimulus, that switched it on or off is removed).

The model that we consider in these exercises appeared originally in the the paper "Construction of a genetic toggle switch in *Escherichia coli*" (Nature vol. 403, January 2000), by Gardner et al.

5. Two types of genes, lets call them type I and type II, are implanted into a **bacteriophage**, which is a virus that infects a specific bacterium—in this case, the bacterium is *E. coli*. Denote the concentrations of genes I and II within the bacteriophage by U and V, respectively.

The interaction of these two genes may be modeled by the following pair of differential equations—which we call the (GT) equations, for "genetic toggle:"

$$\frac{dU}{dt} = \frac{a}{1 + V^\beta} - bU, \qquad \frac{dV}{dt} = \frac{c}{1 + kU^\gamma} - eV. \qquad \text{(GT)}$$

Here, a, b, c, e, k, β, and γ are positive parameters. Also, t is in hours, while U and V are in micrograms per milliliter (μg/mL).

Fill in the each of the blanks below with one of the following terms:

       slowly   quickly   larger   smaller   proportional   inhibits

(a) Consider the first term, $a/(1 + V^\beta)$, in the first differential equation of (GT). Note that, as V gets larger, $1 + V^\beta$ gets _____, which means $a/(1 + V^\beta)$ gets _____, which means (by the first of the (GT) equations) that dU/dt gets _____, which means U grows more _____.
Conclusion: the term $a/(1 + V^\beta)$ represents the fact that gene II _____ production of gene I.

(b) The term $-bU$, in the first of our two differential equations, represents degradation, or dilution, of gene I. This term tells us that this dilution occurs at a rate that is _____ to the concentration of gene I.

(c) Consider the first term, $c/(1 + kU^\gamma)$, in the second differential equation of (GT). Note that, as U gets larger, $1 + kU^\gamma$ gets _____, which means $c/(1 + kU^\gamma)$ gets _____, which means (by the second of the (GT) equations) that dV/dt gets _____, which means V grows more _____.
Conclusion: the term $c/(1 + kU^\gamma)$ represents the fact that gene I _____ production of gene II.

(d) The last term, $-eV$, represents degradation or dilution of gene II. In particular, this term tells us that this dilution occurs at a rate that is _____ to the concentration of gene II.

(e) Because gene II _____ the growth of gene I, and vice versa, these two genes are called "repressor genes."

6. If the feedback of genes I and II on each other is such that the concentration V becomes *larger* than U, the result will be synthesis of a protein called **GFP**, which stands for **green flourescent protein**. When this synthesis is happening, we will say that our genetic toggle switch is **on**. Otherwise, we will say it is **off**. So: the switch being on, or off, corresponds to V being larger than U, or less than or equal to U, respectively. (The flourescence facilitates observing the state of the switch.)
Fill in the blanks, again using the terms from Exercise 5 above:

The smaller the parameter k is, in the above (GT) equation for dV/dt, the _____ the denominator $1 + kU^\gamma$ is, and therefore, the _____ $c/(1 + kU^\gamma)$ is, and therefore, the _____ dV/dt is. So, if we want to make V grow more quickly, we might consider smaller values of k.

We now put this idea into action.

7. Create a Sage program that will use Euler's method to solve the initial value problem consisting of:

- The differential equations (GT); and
- The initial conditions $U(0) = V(0) = 0.1$.

You will need to specify, in your code, the following parameter values:

$$a = 156.25; \quad b = 1; \quad c = 15.6; \quad e = 1; \quad k = 1; \quad \beta = 2.5; \quad \gamma = 1.$$

Also use a starting t-value of $t = 0$ and an ending value of $t = 8$. Now run your program.

(a) With these parameter values and initial conditions, will GFP be produced or not? Please explain.

(b) Now we're going to simulate addition of a chemical "inducer" called IPTG to the solution. Addition of IPTG has the effect of *decreasing* the parameter k. Let's suppose the new value of k is $k = 2.15672 \cdot 10^{-10}$. Change your code to reflect this new value of k, and run the program again. What do you notice this time? Does the toggle switch turn on (eventually) or not?

8. Recall that we would like out switch to be *bistable*, meaning in particular: if we turn our switch on through addition of our inducer IPTG, and then *remove* the IPTG, our switch should remain on.

Let's check this, using our Sage code. Here's how: take your code from part (b), with k specified, at the outset, to be equal to $2.15672 \cdot 10^{-10}$. Then, right after the line

```
while t < tfin :
```
that begins your loop, add the following lines:

```
if t > 2 :
    k=1
```

As usual, the indentation is important. Your program should now look something like this:

```
while t < tfin :
    if t > 2 :
        k = 1
```

This new code has the effect of specifying the following: for the first two hours, we have $k = 2.15672 \cdot 10^{-10}$, meaning the solution contains ITPG. But then, at $t = 2\,\text{h}$, we *remove* the IPTG, meaning we now have $k = 1$, as in Exercise 7(a) above.

(a) Run your new code, and comment on the stability of the switch (once it's in the "on" position).
(b) Replace your above new line of code
```
if t > 2 :
```
with
```
if t > 1 :
```
What happened? What do you think this says about stability?

3.5.3.4 Part 4: Monomers, Dimers, and Trimers
The following exercises require no programming.

A beaker contains three types of molecules, called monomers, dimers, and trimers. We use M, D, and T to stand for the quantities of each of the three respective types. Suppose these quantities are changing over time, according to the following rate equations:

$$M' = -4M^2 - 0.8MD,$$
$$D' = \ \ 2M^2 - 0.8MD,$$
$$T' = \ \ \ \ \ \ \ \ \ \ 0.8MD.$$

Let's suppose that, *initially, there are equal* (nonzero) *quantities of monomers and dimers.*

9. Is D initially increasing or decreasing? (Hint: use the fact, just stated, that $M = D$ initially.) Please explain.
10. At a certain point in the process, D changes from increasing to decreasing (if D is initially increasing), or from decreasing to increasing (if D is initially decreasing). At the point where this happens, what is the value of the ratio M/D? Please explain. (We might call this value the "threshold value" of M/D.)
11. Which of the following four graphs could possibly be a graph of the quantities M, D, and T modeled by the above rate equations? Please explain your reasoning carefully, and on the correct graph, label which curve is M, which is D, and which is T. Hint: start by thinking about increase and decrease.

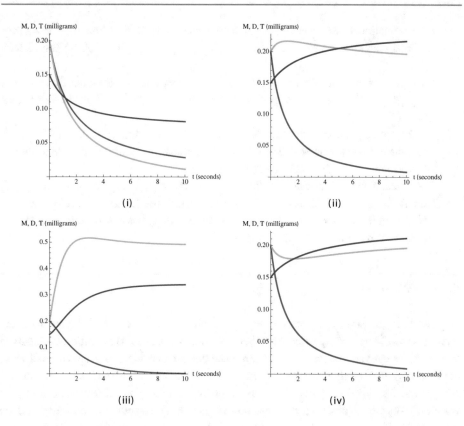

12. Fill in the blanks (try to answer based primarily on quantitative reasoning and mathematics; you shouldn't need any specific knowledge of chemical reactions):

A monomer may react with another monomer to form a dimer. These monomer-to-
monomer reactions cause a decrease in the total quantity of _____. Moreover, The rate at which this occurs is proportional to M^2 (since each of the M milligrams of monomers present has roughly M milligrams of other _____
with which to react). The monomer-to-monomer reactions therefore correspond to the
term _____ in the above equation for M'.

Further, whenever two monomers are lost to a monomer-to-monomer reaction, one
_____ is gained. That is: the rate at which dimers are gained from such reactions
equals half the rate at which _____ are lost to these reactions. Since half
of $4M^2$ equals _____, the monomer-to-monomer reactions account for the term
_____ in the above equation for D'.

A monomer may also react with a dimer to form a _____. The rate at which this
occurs is proportional to the product of the quantity of monomers and the quantity of
dimers (since each of the _____ milligrams of monomers present has _____
milligrams of dimers with which to react). The decrease in M resulting from these
monomer-to-dimer reactions therefore corresponds to the term _____ in the

above equation for M'. Analogously, the decrease in D resulting from these monomer-to-dimer reactions corresponds to the term _____ in the above equation for D'.

Finally, when a monomer and a dimer are lost to a monomer-to-dimer reaction, one _____ is gained. This accounts for the term _____ in the above equation for T'.

13. Use the rate equations at the start of this exercise to compute $M' + 2D' + 3T'$. What does this tell you about $M + 2D + 3T$? How would you interpret this result in terms of the chemical reactions taking place?

14. Show that, in the situation at hand (that is, for the rate equations given at the start of this exercise), the ratio M/D is *always* decreasing. Hint: use the quotient rule to express $(M/D)'$ in terms of M, D, M', and D'; then use the given rate equations to rewrite your result in terms of M and D only.

3.6   Summary

A great variety of natural phenomena may be modeled by **initial value problems**, meaning one or more **differential equations** together with one or more **initial conditions**. Already in this book, we have encountered a number of such phenomena, including: spread of disease, population growth (single and dual species, bounded and unbounded growth, and so on), radiation and radioactive decay, atmospheric pressure, dissolving substances, cooling (or warming) substances, circadian rhythms, neural impulses, fermentation, and genetic toggle switches.

Sometimes, we can find **closed-form solutions** to these initial value problems, meaning solutions of the form

$$Q(t) = \text{some explicit mathematical expression in } t$$

for each of the quantities $Q(t)$ encompassed by the model. For example, we were able to do so in the case of the exponential growth and decay initial value problems of Sect. 3.1, as well as the logistic growth problem given by the differential equation (3.33) and the initial condition $R(0) = R_0$. We were also able to do this for the "Newton's Law of Cooling" problem of Exercise 24, Sect. 1.7.4, and Exercise 12, Sect. 3.1.5.

Closed-form solutions to initial value problems are quite useful, in that they give us very explicit information, which can, in principle, be studied in detail locally (meaning at or near a given point), as well as globally (meaning over large ranges of the independent variable).

However, closed-form solutions can be difficult or impossible to find. In fact, they are the exception rather than the rule. Fortunately, though, when such solutions are unavailable, we can still use **Euler's method**, which is based on the idea that

$$\text{new } Q = \text{old } Q + \Delta Q \approx \text{old } Q + Q' \Delta t.$$

This idea, together with given initial conditions, can be used to generate a *list*

$$Q(t_0), \; Q(t_0 + \Delta t), \; Q(t_0 + 2\Delta t), \; Q(t_0 + 3\Delta t), \ldots$$

of successive (approximate) values of Q, for each of the quantities Q under consideration. (Here, t_0 is the instant at which things "start.") And from such lists, one can obtain graphs.

We applied Euler's method quite explicitly to the SIR problem of Chap. 1 (for which no closed-form solution is known to exist). Euler's method was also used (without explicit mention, until now) to generate a number of the graphs appearing earlier in this chapter, and was invoked explicitly in various of the exercises from Sects. 3.4 and 3.5.

In the next chapter, we'll develop the notion of **integration**, which will provide us with additional tools for finding *closed-form* solutions to a variety of differential equations and initial value problems, of **pure-time**, **autonomous**, and **mixed** types. (See the beginning of Sect. 4.8 for the definitions of these terms.)

Integration

There are many contexts—energy, work, force, profit/loss, area, volume, distance traveled, and mass are just a few—where the quantity in which we are interested equals some other quantity times elapsed time, or times some change in an independent variable. For example, the electrical **energy** needed to burn a $100\,\mathrm{W}$ (watts) light bulb for T hours is $100T$ watt-hours, which is the product of the $100\,\mathrm{W}$ **power** demand times the elapsed time T. As another example, if a uniform **pressure** of $25\,\mathrm{lbs}$ per square inch is applied to an object of area A square inches, then the total **force** applied to that object is $25A$ pounds.

The calculation becomes more complicated, though, if lights are turned off and on during the time interval of length T, or if the applied pressure varies across the surface of the object. We face the same complication in any of the contexts delineated in the previous paragraph, when quantities are varying. To handle such complications, we will introduce the **integral**.

As we will see, the integral itself can be viewed as a function of the independent variable. By analyzing the rate of change of that function, we will find that every integral can be expressed as the solution to a particular initial value problem. This observation will allow us to evaluate many integrals through a process known as **antidifferentiation**.

We will further see how antidifferentiation can, itself, be applied to the solution of a variety of differential equations and initial value problems.

4.1 Power and Energy

The work done by electricity is usually referred to as **(electrical) energy**.

The rate at which electrical energy is supplied, or consumed, is referred to as **power**. In this section, we explore relationships between power and energy. The broad purpose of this exploration is to introduce the idea of an **accumulation function,** which generalizes the notion of a product to situations where one of the factors varies.

© The Author(s), under exclusive license to Springer Nature Switzerland AG 2023
E. Stade and E. Stade, *Calculus: A Modeling and Computational Thinking Approach*,
Synthesis Lectures on Mathematics & Statistics,
https://doi.org/10.1007/978-3-031-24681-4_4

Ultimately, our goal is to understand how accumulation functions are related to derivatives, and to *areas*.

4.1.1 Part A: Power Supplied at a Constant Rate

Suppose electrical energy is supplied, or consumed, at a **constant** rate over a given period of time. Again, this rate is called *power*. So in this situation, we have the formula

$$\text{energy} = \text{power} \cdot \text{elapsed time}, \tag{4.1}$$

where "elapsed time" refers to duration of the interval in question.

Example 4.1.1 A 60 W bulb burning for two and a half hours consumes

$$60 \text{ W} \cdot 2.5 \text{ h} = 150 \text{ watt-hours}$$

of energy.

Note that we have used **watts**, also denoted W, as our units for power, and hours (h) as our units for time, whence our units for energy become watt-hours, or Wh.

Keep in mind that power is a **rate**. It might be helpful to think of a watt as a "watt-hour per hour;" the word "per" helps remind us of rates of change.

Other units for electric power are the kilowatt (kW) (=1,000 W), the megawatt (mW) (=1,000,000 W), and the gigawatt (gW) (=1,000,000,000 W). So electrical *energy* is also measured in kilowatt-hours (kWh), in megawatt-hours (mWh), and in gigawatt-hours (gWh).

Energy is also sometimes measured in joules. By definition, 1 J equals one thirty-six-hundredth of a watt-hour. Or, since there are 3600 s in an hour,

$$1 \text{ J} = \frac{1}{3600} \text{ watt-hours} \cdot 3600 \, \frac{\text{s}}{\text{h}} = 1 \text{ watt-second.}$$

To fully appreciate the relationship between energy and power, it will be instructive to revisit the light bulb of Example 4.1.1 above. But this time, we will consider cumulative energy consumption not only at the end of the 2.5 h time period, but at any time T "along the way."

Example 4.1.2 What is the cumulative amount of energy consumed, call it $E(T)$, by the bulb of Example 4.1.1, from the beginning of the 2.5-hour period to time T, where T is any number of hours between 0 and 2.5?

Solution. By (4.1), the amount of energy consumed over the given interval of time is

$$E(T) = 60 \text{ W} \cdot T \text{ h} = 60T \text{ watt-hours.}$$

We've used a "T" rather than a "t" in this example to emphasize the fact that we are measuring a cumulative effect over time. (Capital letters seem somehow more cumulative than lower-case ones.)

To conclude this subsection, let's look more closely at the relationship between the above energy function $E(T)$ and the power demand $p(T)$ at time T, where T is, again, any number between 0 and 2.5. Of course, power demand is constant over this interval; we have $p(T) = 60$ (watts) for any such T. Fig. 4.1 shows the graphs of $p(T)$ and $E(T)$ over this period.

(This time, we're using a capital "T" for both power and energy, so that we may compare the two quantities more readily.) The important thing to note from Fig. 4.1 is that **the slope of the energy function** $E(T)$ **equals the height of the power function** $p(T)$—both of these numbers equal 60 W. But the height of the power function is just $p(T)$ (which is constant since power is being supplied at a constant rate), and the slope of the energy function is just $E'(T)$ (since the slope of a linear function equals the derivative of that function). So what we have just seen is that

$$E'(T) = p(T). \tag{4.2}$$

Of course, these arguments apply, so far, only to a situation where power is supplied at a constant rate. But similar conclusions may be drawn in more general situations. We will see this over the remainder of this section (and chapter).

4.1.2 Part B: Power that Varies in Steps

Consider the following modification of Example 4.1.1.

Example 4.1.3 A 3-way light bulb burns at 60 W for 3.5 h, at 100 W for another 1.5 h, and at 45 W for another 3 h. How much cumulative energy is consumed by the light bulb over the entire 8-hour period?

The key here is to use Eq. (4.1) in *stages*, and then to add up the results from those stages. That is: over the first 3.5 h, 60 W · 3.5 h = 210 watt-hours of energy are consumed; over the next 1.5 h, 100 W · 1.5 h = 150 watt-hours of energy are consumed; over the next 3 h, 45

$W \cdot 3 \, h = 135$ watt-hours of energy are consumed. So the total energy consumption, over the 8-hour period, is given by

$$\text{energy} = 210 + 150 + 135 = 495 \text{ watt-hours.}$$

Now, as we did in Part A of this section, let's ask a more general question.

Example 4.1.4 What is the cumulative energy consumption $E(T)$ of the above 3-way light bulb, in watt-hours, T hours from the beginning of our eight-hour period?

Solution. The easiest way to do this is to first plot a few points on our graph of $E(T)$. We use the following information.

1. At time $T = 0$, there is no energy consumption. So $E(0) = 0$.
2. In Example 4.1.3 above, we saw that 210 watt-hours are consumed over the first 3.5 h. So $E(3.5) = 210$.
3. In that example we also saw that, over the next 1.5 h, 150 watt-hours are consumed. So total energy consumption over the first five hours is $E(5) = 210 + 150 = 360$.
4. Finally, we computed in that example that the total energy consumption over the entire eight hours is $E(8) = 495$ watt-hours.

So our graph of $E(T)$ goes through the points $(0, 0)$, $(3.5, 210)$, $(5, 360)$, and $(8, 495)$. Moreover, between any two of these points that are adjacent, the graph of $E(T)$ is *linear*. Why? For the same reason as in Part A above: if power is constant on an interval, then energy increases at a constant rate on that interval.

A graph of this function $E(T)$ appears in Fig. 4.2. Of course, a graph is not a formula. But a formula for $E(T)$ is readily obtained, using either the point-slope or the two-point form of the equation for a line. For example: the line segment corresponding to $5 \leq T \leq 8$ passes through $(5, 360)$ and has slope (power rating) equal to 45, so by the point-slope form, it has equation

$$E(T) - 360 = 45(T - 5),$$

or, solving for $E(T)$, $E(T) = 45T + 135$.

Through similar calculations for the other two intervals, we ultimately obtain

$$E(T) = \begin{cases} 60T & \text{if } 0 \leq T \leq 3.5, \\ 100T - 140 & \text{if } 3.5 \leq T \leq 5, \\ 45T + 135 & \text{if } 5 \leq T \leq 8. \end{cases} \tag{4.3}$$

As was the case for our 60 W bulb of Part A of this section, we see here that, for our 3-way bulb, **the slope of the energy function $E(T)$ equals the height of the power function $p(T)$.**

Actually, this is not *quite* true in the present case. The energy function $E(T)$ in Fig. 4.2 does not in fact *have* a slope at the T-values $T = 3.5$ and $T = 5$. (At these points, $E(T)$ has "corners," and is therefore not locally linear.) But at any other point in between 0 and 8, we see that the slope of the $E(T)$ graph *does* equal the height of the $p(T)$ graph. (Between 0 and 3.5, the slope of $E(T)$ and the height of $p(T)$ are both equal to 60; between 3.5 and 5, this slope and this height are both equal to 100; between 5 and 8, these are both equal to 45.) In other words, except at the points where $E(T)$ is not locally linear, we have, again,

$$E'(T) = p(T). \tag{4.4}$$

We next consider an even more general situation.

4.1.3   Part C: Power that Varies Continuously

Suppose that power demand $p(t)$ of a certain town, over a 24 h period—from midnight, denoted $t = 0$, to midnight, denoted $t = 24$—is described by Fig. 4.3.

This graph is analogous to the graphs on the left-hand sides of Figs. 4.1 and 4.2. Of course, the situation in Fig. 4.3 is more complicated, in that, here, power is neither constant, nor constant "in steps;" instead, $p(t)$ is continuously varying with t. Still we wish to answer, at least approximately, the same kinds of questions as were posed in Parts A and B of this section.

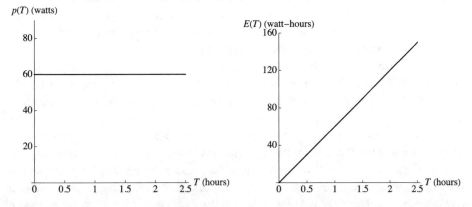

Fig. 4.1 Power and energy, for an interval over which power demand is constant

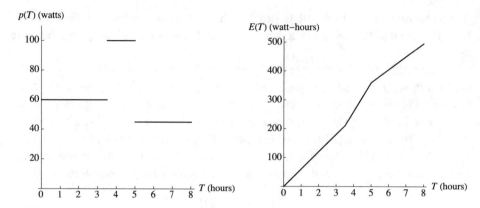

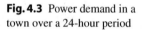

Fig. 4.2 Power and energy, when power demand is "constant in steps"

Fig. 4.3 Power demand in a
town over a 24-hour period

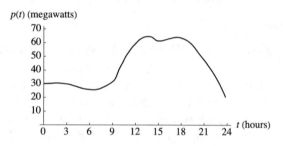

Example 4.1.5 What, at least approximately, is the total amount of energy consumed by
the above town, over the given 24-hour period?

Solution. They key is to break the interval [0, 24] into smaller subintervals, such that, on
each of these subintervals, the power function $p(t)$ does not vary "too much." On each of
these intervals, we can then **approximate** energy consumption using (4.1). Adding together
all of the energy consumption numbers obtained in this way, we can deduce an estimate for
the total, cumulative energy usage over the entire period.

   To illustrate, consider Fig. 4.4, where we have divided the 24-hour period of Fig. 4.3 into
nine different subintervals, such that $p(t)$ does not vary too much on any one of these subintervals. And we have drawn a rectangle over each subinterval, for reasons to be explained
shortly.

   Note that the subintervals have varying lengths. The idea here is this. In places where
its graph is relatively flat (horizontal), $p(t)$ remains "roughly constant" for relatively long
periods of time. So we can cover such parts of the graph with relatively long subintervals.
Conversely, where its graph is steep, $p(t)$ changes rapidly. So, in such places, we need
relatively short subintervals, to assure that $p(t)$ does not change too much on any one of
those subintervals.

Fig. 4.4 Breaking [0, 24] into subintervals where $p(t)$ is "roughly constant"

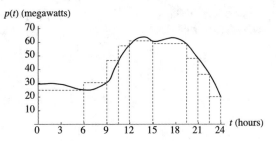

Since $p(t)$ is roughly constant on each subinterval, we can approximate $p(t)$, on any subinterval, by its value at *any point*—to be concrete, let's say the *right endpoint*—of that subinterval. (We could have chosen left endpoints, or midpoints, etc., just as well. Or we could have chosen a different kind of point on each subinterval. We'll say more about this choice soon.) In other words, we can imagine that $p(t)$ is "constant in steps," and that the heights of the dashed, horizontal lines in Fig. 4.4 give the values of $p(t)$ on the various steps.

Now the first of the subintervals in Fig. 4.4 has power value of about 25 (we read this value off of the graph as well as we can), and length 6. Therefore, the energy consumption over this period is roughly 25 mW · 6 h = 150 mWh. Note, incidentally, that 25 mW is the **height** of the first (that is, the leftmost) dashed rectangle in Fig. 4.4, and 6 h is the **baselength** of this rectangle. So 25 mW · 6 h = 150 mWh is the **area** of this first rectangle.

The second interval in Fig. 4.4 has power value of about 30 and length 3, so the energy consumption over this period is about 30 mW · 3 h = 90 mWh. And note that 30 mW · 3 h = 90mWh is the **area** of the second dashed rectangle in Fig. 4.4.

And so on up through the ninth subinterval. The cumulative amount of energy consumed is the sum of amounts consumed over all nine subintervals. And, as we indicated directly above, the energy consumed over any one of these intervals is given, approximately, by the **area** of the dashed rectangle drawn over that interval.

Reading the appropriate power values off of the graph, then, we find that

cumulative energy consumed over the 24-hour period

$=$ sum of energy amounts over each of the smaller subintervals

$\approx$ sum of areas of the dashed rectangles

$\approx 25 \cdot 6 + 30 \cdot 3 + 47 \cdot 1.5 + 57 \cdot 1.5 + 61 \cdot 3 + 59 \cdot 4.5 + 48 \cdot 1.5 + 36 \cdot 1.5 + 20 \cdot 1.5$

$= 1000.5$ megawatt-hours.   (4.5)

Note that the cumulative energy value obtained above is only an *estimate*. How might we get a better estimate? The answer is clear: start with a "step function" (that is, a function that is constant in steps) that approximates the power graph *more closely*. In principle, we

Fig. 4.5 Subdividing [0, 24] into narrower subintervals

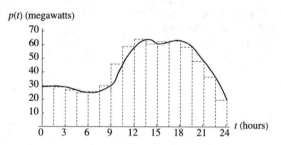

$p(t)$ (megawatts)

can get as good an approximation as we might desire this way. We are limited only by the precision of the power graph itself.

For example, consider the following refinement of the subdivision shown in Fig. 4.4. Here, all subintervals have the same length, equal to 1.5 h. And we have, again, "sampled" $p(t)$, on each subinterval, at the right endpoint of that subinterval. As before, the energy consumed over a given subinterval is approximately equal to the **height** (the sampled value of the power function), times the **baselength** (the elapsed time), of the dashed rectangle over that interval. In other words (again): the energy consumed over a given subinterval is approximately equal to the **area** of the dashed rectangle over that interval (Fig. 4.5).

From this picture, we might obtain the following approximation:

cumulative energy over the 24-hour period

$= $ sum of energy amounts over each of the smaller subintervals

$\approx $ sum of areas of the dashed rectangles

$\approx 30 \cdot 1.5 + 29 \cdot 1.5 + 27 \cdot 1.5 + 25 \cdot 1.5 + 25 \cdot 1.5 + 31 \cdot 1.5 + 46 \cdot 1.5 + 58 \cdot 1.5$

$+ 63 \cdot 1.5 + 60 \cdot 1.5 + 61 \cdot 1.5 + 62 \cdot 1.5 + 58 \cdot 1.5 + 48 \cdot 1.5 + 36 \cdot 1.5 + 20 \cdot 1.5$

$= 998.5 \quad$ megawatt-hours. $\hfill (4.6)$

In summary, we can determine the energy consumption of the town by a sequence of successive approximations. The steps in the sequence are listed in the box below.

Step 1. Approximate the power demand by a step function.
Step 2. Estimate energy consumption from this approximation. Note that this approximation is a sum of areas of rectangles defined by the step function.
Step 3. Improve the energy estimate by choosing a new step function that follows power demand more closely.

Steps for computing energy consumption from a graph of the power function

Now, as we did in Parts A and B above, let's look at the energy consumed as a function of time.

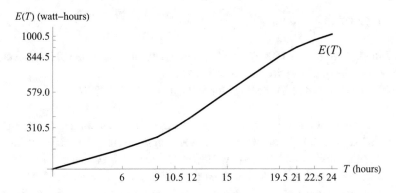

Fig. 4.6 An approximation to the energy function corresponding to Fig. 4.3

Example 4.1.6 Use Fig. 4.4 to obtain an approximate graph of the energy function $E(T)$ for this town, for $0 \leq T \leq 24$.

Solution. Having approximated $p(t)$ by a step function in Fig. 4.4, we can now follow essentially the method of Example 4.1.4 to find an approximation to $E(T)$. This approximation will be linear in steps, like the graph of $E(T)$ on the right-hand side of Fig. 4.2, because our approximation to $p(t)$ is constant in steps.

We being by noting that $E(0) = 0$. We also have $E(6) \approx 25 \cdot 6 = 150$, since $p(t) \approx 25$ over the first interval of length 6 h. Over the next 3 h, we estimated that $p(t) \approx 30$, so the energy consumed over that 3-hour interval is $\approx 30 \cdot 3 = 90$. So the total energy consumed over the first 9 h is $\approx 150 + 90 = 240$; that is, $E(9) \approx 240$. And so on, all the way up to our last value $E(24) \approx 1000.5$. (This is what we found in Example 4.1.5 above.) So our graph of $E(T)$ passes through the points $(0, 0)$, $(6, 150)$, $(9, 240)$, ...$(24, 1000.5)$. We plot these points in the plane, and connect them with line segments. The result is Fig. 4.6.

(A better approximation to $E(T)$ could be obtained by starting with Fig. 4.5, instead of Fig. 4.4.)

We could also find a mathematical *formula* for our approximation to $E(T)$, using the method of Example 4.1.4. And using this formula, we could verify that $E'(T) \approx p(T)$ at all points on $[0, 24]$ where $E'(T)$ exists. The computations here are a bit arduous, though, and are not particularly illuminating, so we skip them.

We call E an **accumulation function** for p because energy consumed over an interval is a cumulative effect of power consumption during that interval. In the next section, we'll expand on the notion of accumulation function, and will see that this notion is intimately related to area, and to *derivatives*.

4.1.4   Exercises

1. On Monday evening, a 1500 W space heater is left on from 7 until 11 pm. How many kilowatt-hours of electricity does it consume?
2. Suppose a space heater has settings for 500, 1,000, and 1, 500 W.

   (a) On Tuesday, we put this space heater on the 1,000 W setting from 6 to 8 pm, then switch it to 1, 500 W from 8 till 11 pm, and then switch it to the 500 W setting through the night until 8 am, Wednesday. How much energy is consumed, in kWh, by this heater over this $2 + 3 + 9 = 14$ hour period?
   (b) Sketch the graphs of power demand $p(T)$ and accumulated energy consumption $E(T)$ for the space heater from 6 pm Tuesday evening to 8 am Wednesday morning. Determine whether $E'(T) = p(T)$ (wherever $E'(T)$ exists) in this case.

3. If energy is consumed (by an electronic device, a household, a town, etc.) over a period of time, then we define the **average power demand** over this period by:

$$\text{average power demand} = \frac{\text{total energy consumption}}{\text{elapsed time}}.$$

   (a) What is the average power demand of the space heater in Exercise 2(a) above, over the 14-hour period in question? What are your units?
   (b) Figures 4.4 and 4.5 depict the power demand of a town over a 24-hour period. Give two different estimates of the **average** power demand of the town during that period. Explain how you got your estimates.

4. The graph below depicts power generation, through rooftop solar panels, and power consumption at an elementary school in Boulder, Colorado. The green, bell-shaped curve, call it $p_g(t)$, represents power **generated** (that is, into the grid) from the solar panels. The red, somewhat jagged curve, call it $p_c(t)$, represents power **consumption** (that is, from the grid) at the school.
   The vertical axis is the power axis. The range is 0–10 kilowatts (kW). The horizontal axis is the time axis. The domain is from 12 AM Monday, March 18 until 12 AM Tuesday, March 19, 2013. Each of the ticks on the time axis represents 1/2 h. (The bolder ticks are spaced 3 h apart.)

   (a) Make a copy of this graph. On your copy, dash in a rectangle over each of the intervals $[0, 2]$, $[2, 4]$, $[4, 6]$, ..., $[22, 24]$. The height of the rectangle over each interval should be the value of the power **generated** function $p_g(t)$ (again, in green) at the *right* endpoint of that interval. (Some of your rectangles will have height zero.)
   (b) Let $E_g(T)$ denote cumulative energy **generated** by the solar cells, from time 0 to time T. Use your above rectangles to estimate $E_g(2)$, $E_g(4)$, $E_g(6)$, ..., $E_g(24)$.

(Careful: $E_g(T)$ measures *cumulative* energy generated. So, for example, calculation of $E_g(10)$ will involve the rectangles over [0, 2], [2, 4], [4, 6], [6, 8], and [8, 10], not just the rectangle over [8, 10].) What are the units for $E_g(T)$?

(c) Repeat parts (a) and (b) of this exercise, except this time use the power **consumed** curve $p_c(t)$ (again, in red) to estimate the energy **consumed** function $E_c(T)$, again at $T = 0, 2, 4, \ldots 24$. (Use a separate copy of the graph below to draw your rectangles.)

(d) On a separate graph, with time on the horizontal axis and *energy* on the vertical axis, plot the points

$$(0, E_g(0)), \ (2, E_g(2)), \ (4, E_g(4)) \ldots, (24, E_g(24))$$

(where you are using your estimates from part (b) of this exercise as approximations to E_g). Connect the dots with line segments, to sketch a rough approximation to $E_g(T)$.

(e) On the *same set of axes* as you used in part (d) of this exercise, plot the points

$$(0, E_c(0)), \ (2, E_c(2)), \ (4, E_c(4)) \ldots, (24, E_c(24)),$$

using your estimates from part (c) of this exercise as your approximations to E_c.

(f) Do your graphs of $E_g(T)$ and $E_c(T)$ ever intersect? If so, where? What's the significance of this?

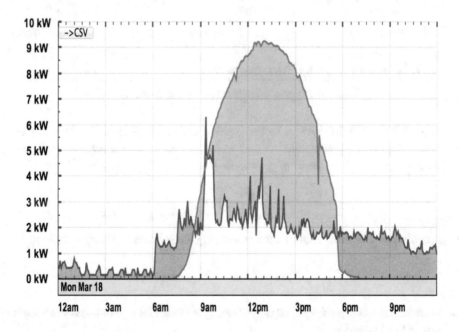

4.2   Accumulation Functions and Definite Integrals

We begin with a definition that generalizes the relationship between power and energy examined in the previous section. But, in a slight departure from the terminology of that section, we formulate our definition in terms of **net changes** in E and t. This is done so that, ultimately, we may apply the full power of calculus (which is, again, the the mathematics of change) to the study of accumulation functions.

We also introduce some new notation for accumulation functions.

---

1. Suppose $E(t)$ and $p(t)$ are measurable quantities, dependent on a variable t, and that

$$\Delta E = p(t)\Delta t, \tag{4.7}$$

over any interval of t-values where $p(t)$ is constant. (Here, Δt is the length of the interval in question, ΔE is the change in E over that interval, and t is any point in the interval. Since $p(t)$ is constant on this interval, it does not matter which point t in that interval we choose.)

Then we say that E is an **accumulation function** for p.

2. If E is an accumulation function for p on an interval $[a, b]$, then we define **the definite integral of** p **over** $[a, b]$, written

$$\int_a^b p(t)\,dt, \tag{4.8}$$

to be the net change in E over this interval.

Definition of accumulation function and of definite integral

Example 4.2.1 (a) In the previous section, we saw that energy is an accumulation function for power. So, given a power function $p(t)$ on an interval $[a, b]$, the net change in energy E—that is, the energy consumed or generated—over that interval may be written

$$\int_a^b p(t)\,dt.$$

In particular, we saw in Example 4.1.5 that, for $p(t)$ the function of Fig. 4.3, we have

$$\int_0^{24} p(t)\,dt \approx 1000.5 \quad \text{megawatt-hours.}$$

(A more refined estimate, developed in Eq. (4.6), gave 998.5 megawatt-hours as an approximation to this integral.)

In Example 4.1.6 above, we sketched a graph of $E(T)$. But we claim that, in fact,

$$E(T) = \int_0^T p(t)\, dt. \tag{4.9}$$

Why is this claim true? Well on the one hand, the quantity $E(T)$ on the left-hand side of (4.9) equals $E(T) - E(0)$, since $E(0) = 0$ (no energy has been consumed at the outset). On the other hand, the right-hand side of (4.9), by the above definition of the definite integal, equals the change in E from $t = 0$ to $t = T$, and this change is just the difference $E(T) - E(0)$. So the two sides of (4.9) are indeed equal.

In other words: in Example 4.1.6, we are considering the integral

$$\int_0^T p(t)\, dt$$

as a function of T.

(b) Distance equals velocity times time, or in other words

change Δd in displacement (distance) $=$ velocity $v(t)$ times elapsed time Δt,

as long as velocity is constant over the interval under consideration. So distance is an accumulation function for velocity.

So an automobile traveling with velocity $v(t)$ miles per hour, from time $t = 0$ to $t = 5$ (hours), travels a distance of

$$\int_0^5 v(t)\, dt \quad \text{miles.}$$

We'll consider an example of such a journey later in this section.

Here is a short table of accumulation functions, including those considered in the above example. Can you think of others?

E	p	t
energy	power	time
distance	velocity	time
work	force	distance
mass	density	length
force	pressure	area
area	height	length
number of Ebola deaths	death rate	time

Note that the independent variable need not be time.

The goal of this and the next few sections will be the **evaluation** of accumulation functions, or more precisely of definite integrals $\int_a^b p(t)\,dt$, given information about the underlying quantity p that is "accumulating." In the present section, we'll evaluate such integrals *approximately*, using essentially the ideas of the previous section, where E denoted energy and p denoted power.

Remark 4.2.1 In considering accumulation functions, and integrals $\int_a^b p(t)\,dt$, we will always take it for granted that the function $p(t)$ being "accumulated," or "integrated," is not too pathological on the interval $[a, b]$ in question. This will ensure that the results we derive, often somewhat informally, are in fact valid and may be proved rigorously.

In particular, it's enough to assume, throughout, that $p(t)$ is *piecewise continuous* on the interval $[a, b]$ under consideration. This means that $[a, b]$ can be subdivided into finitely many closed, bounded subintervals, on each of which f is continuous.

The kinds of functions that we encounter in context are typically piecewise continuous. For example, if f is continuous on $[a, b]$, then it's piecewise continuous there. Also, "step functions," like the one in the left-hand graph of Fig. 4.2, or like the ones defined by the tops of the dashed rectangles in Figs. 4.4 and 4.5, are piecewise continuous. And so on.

4.2.1   Evaluation of Definite Integrals, Part A: Integrals and Area

Let E be an accumulation function for a function p on an interval $[a, b]$. For the moment, we'll assume that $p(t)$ is nonnegative on $[a, b]$, which is to say that the graph of $p(t)$ never dips below the t-axis on this interval. Later on, we'll relax this assumption.

We should not expect the equation

$$\Delta E = p(t)\Delta t \tag{4.10}$$

to hold over the entire span of the interval $[a, b]$ (where $\Delta t = b - a$), since $p(t)$ will not, in general, be constant over this interval. However, if we choose a *subinterval* of $[a, b]$, and the length Δt of this subinterval is small enough, then we would expect $p(t)$ not to vary too much over this subinterval—that is, we would expect $p(t)$ to be "approximately constant" there. So we would expect, in turn, that Eq. (4.10) is "approximately true," meaning

$$\Delta E \approx p(t)\Delta t, \tag{4.11}$$

on this interval. Here t could be any point in the interval; since $p(t)$ doesn't vary much on the subinterval, the right-hand side of (4.11) shouldn't vary much with our choice of t (as long as t belongs to the interval in question).

The situation just described is exemplified by Fig. 4.7.

Consider the dashed rectangle shown in Fig. 4.7. Its baselength is Δt and its height is $p(t)$, where t is a point in the given interval of length Δt. The area of this rectangle is, of

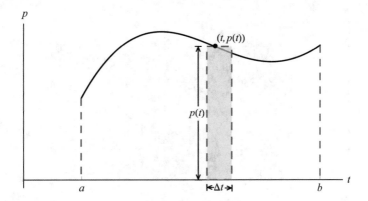

Fig. 4.7 A function $p(t)$ and a "representative rectangle"

course, equal to its height times its baselength, so this area equals $p(t)\Delta t$. That is, by (4.11), the area of this rectangle is approximately equal to the change ΔE in the accumulation function $E(t)$, over this interval.

Now imagine that we've spanned the entire interval $[a, b]$ with subintervals like the one above, and have drawn a rectangle like the one above over each of these subintervals. The area of each rectangle approximates, as we've just seen, the change ΔE in the accumulation function over the appropriate interval. So the *sum* of the areas of the rectangles approximates the *sum* of these changes in E. But the sum of the changes in E over the subintervals equals the *overall* change in E over all of $[a, b]$. So the change in E over $[a, b]$ is approximately given by the sum of the areas of these rectangles.

But recall that we have denoted the change in E over $[a, b]$ by the definite integral

$$\int_a^b p(t)\, dt.$$

Conclusion: this integral is approximated by the sum of the areas of the rectangles depicted in Fig. 4.8.

Now what happens as we "refine" Fig. 4.8: that is, we take more and more rectangles, all of which have narrower and narrower baselengths? Well on the one hand, the area of each rectangle should give us a better approximation to the change in E on the interval where that rectangle "lives." So the sum of these areas should give us a better approximation to the change in E over $[a, b]$. On the other hand, the narrower the rectangles, the better they "fit" under the graph of $p(t)$, so the closer the *sum* of their areas is to the actual **area under the graph of $p(t)$, over $[a, b]$**.

To summarize the arguments so far: if $E(t)$ is an accumulation function for $p(t)$, over an interval $[a, b]$, and $p(t) \geq 0$ on that interval, then:

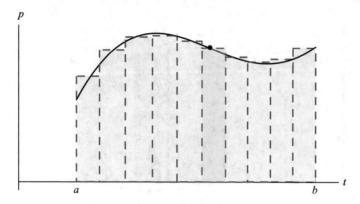

Fig. 4.8 The sum of the areas of the rectangles approximates the change in E over $[a, b]$

change in E over $[a, b]$

= sum of changes in E over subintervals into which $[a, b]$ is subdivided

≈ sum of terms of the form $p(t)\Delta t$, like the one shown in Fig. 4.7

= sum, over all the subintervals, of areas of rectangles like the ones shown in Fig. 4.8

≈ the area under the graph of $p(t)$, over $[a, b]$.   (4.12)

There are two separate types of approximations going on here: (i) the change in E over a subinterval is only *approximately* equal to the product $p(t)\Delta t$ there (because $p(t)$ is not actually constant on such a subinterval); and (ii) the areas of all the rectangles, added together, is only *approximately* equal to the area under the graph (because the "space" filled up by the rectangles is not exactly the same as the "space" under the graph). But both of these approximations should improve as we take more and more, narrower and narrower, rectangles. So as we do this, all of the "≈" symbols in (4.12) should become "=," which means that the far left-hand and right-hand sides must, in the end, be *equal*.

We therefore have the following conclusion.

If $E(t)$ is an accumulation function for $p(t)$, and $p(t) \geq 0$ on $[a, b]$, then the change

$$\int_a^b p(t)\, dt$$

in E over $[a, b]$ equals the area under the graph of $p(t)$, over $[a, b]$.

Accumulation and area

As mentioned earlier, we will soon drop the requirement that $p(t) \geq 0$ on $[a, b]$. Of course, whenever $p(t) < 0$, the graph of $p(t)$ lies *under* the x-axis, so we'll have to reinterpret the phrase "over $[a, b]$" in the above conclusion. See Sect. 4.4.

The arguments that brought us to the above conclusion are somewhat informal. But they can be made rigorous (so that the conclusion really does hold) as long as $p(t)$ is not too "pathological" a function.

Example 4.2.2 During the course of a certain five-hour journey, a car travels with velocity $v(t)$ given by Fig. 4.9 below. Estimate, to within 50 miles, the distance traveled by this car on this trip.

Solution. As noted earlier, distance is an accumulation function for velocity. So by the above conclusion, the distance

$$\int_0^5 v(t)\, dt$$

traveled by the car equals the area under the given graph. To approximate this area, consider the two figures below.

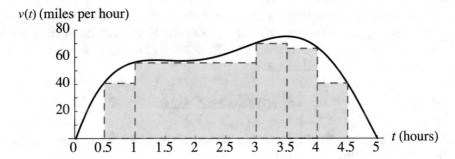

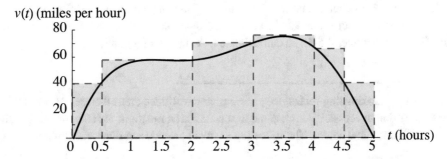

In the top figure, we have rectangles whose areas add up to about

$$40 \cdot 0.5 + 55 \cdot 2 + 70 \cdot 0.5 + 65 \cdot 0.5 + 40 \cdot 0.5 = 217.5,$$

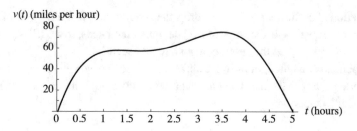

Fig. 4.9 A velocity function

and these rectangles clearly fill up *less* than the total area under the graph. So the area under the graph must be more than 217.5. In the bottom figure, we have rectangles whose areas add up to about

$$40 \cdot 0.5 + 60 \cdot 1.5 + 70 \cdot 1 + 75 \cdot 1 + 65 \cdot 0.5 + 40 \cdot 0.5 = 307.5,$$

and these rectangles clearly fill up *more* than the total area under the graph. So the area under the graph must be less than 307.5.

The above two approximations are 90 miles apart, so we can't use either as our final estimate, if we want to guarantee that we are no more than 50 miles off. However, note that the average of these two approximations is $(217.5 + 307.5)/2 = 262.5$. So we can estimate that our car traveled 262.5 miles; that is,

$$\int_0^5 v(t)\, dt \approx 262.5 \quad \text{miles.}$$

Since neither 217.5 nor 307.5 is more than 50 miles from 262.5, our estimate is within 50 miles of the actual distance traveled.

This assumes, of course, that we were fairly accurate in reading values off of the vertical axis of our graph. It turns out that the actual area under the graph shown above is about 268 (see Exercise 6 of Sect. 4.5.4), so our estimate is, in fact, pretty good.

Of course, approximating rectangles can be used to estimate areas themselves, even when these areas don't necessarily correspond to accumulation functions for other quantities (like energy, distance, and so on). The following example provides an illustration of this.

Example 4.2.3 Approximate the area under the graph of $f(x) = \sqrt{x - 1}$, from $x = 1$ to $x = 6$, using a *left endpoint* approximation with ten subintervals of equal length. By "left endpoint approximation" we mean: use the left endpoint of each subinterval as the sampling point (the point at which the corresponding rectangle "tops out") in that interval (Fig. 4.10).

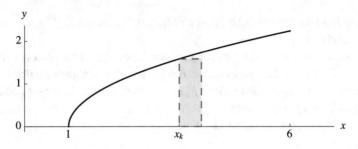

Fig. 4.10 $f(x) = \sqrt{x-1}$ and a representative rectangle

Solution. To keep the picture relatively uncluttered, we've drawn in a single representative rectangle, in Fig. 4.10, rather than all ten of the rectangles that will be used in the approximation.

We've stated that all subinterval lengths should be the same. Let's denote this common length by Δx: we compute that

$$\Delta x = \frac{\text{length of the interval } [1, 6]}{\text{number of subintervals}} = \frac{6-1}{10} = 0.5.$$

Also, we're doing a left endpoint approximation. So our first sampling point, call it x_1, is the left endpoint of our first subinterval; that is, $x_1 = 1$. Our second sampling point x_2 is the left endpoint of the second subinterval; so $x_2 = 1.5$, and so on all the way up to our tenth subinterval, whose left endpoint is $x_{10} = 5.5$.

So our area is approximated by the following sum:

$$\int_1^6 \sqrt{x-1}\,dx \approx f(x_1)\Delta x + f(x_2)\Delta x + \cdots + f(x_{10})\Delta x$$

$$= \Delta x(f(x_1) + f(x_2) + \cdots + f(x_{10}))$$
$$= 0.5(f(1) + f(1.5) + \cdots + f(5.5))$$
$$= 0.5(\sqrt{1-1} + \sqrt{1.5-1} + \cdots + \sqrt{5.5-1})$$
$$= 0.5(0 + 0.707107 + 1 + 1.22474 + 1.41421 + 1.58114 + 1.73205 + 1.87083 + 2 + 2.12132)$$
$$= 6.82570. \tag{4.13}$$

Note that, because we have a *formula* for the function f being "accumulated"—that is, integrated—we can use this formula to determine the heights of the rectangles. That is, we don't need to try and estimate these heights visually from the graph.

Later, we'll see that the true area $\int_1^6 \sqrt{x-1}\,dx$ equals 7.4536 (to four decimal places). So our approximation is an underestimate. This can can also be seen from Fig. 4.10: as the representative rectangle illustrates, all of our left endpoint rectangles will *undershoot* the graph of $f(x)$. So the sum of their areas will be less than the area $\int_1^6 \sqrt{x-1}\, dx$ we are trying to estimate.

More generally, as the above figure suggests, the following is true. On any interval $[a, b]$ where $f(x)$ is *increasing* (and non-negative), any left endpoint approximation to $f(x)$ will underestimate the area under the graph of $f(x)$, over that interval. Analogous things happen when $f(x)$ is *decreasing* on an interval, and/or we use a *right endpoint* approximation. You should think about whether the approximations give underestimates or overestimates in each of these cases.

All of our integration so far has been approximate. In Sect. 4.5 below, we'll see that, in some cases, we can evaluate definite integrals *exactly*, using a process called *antidifferentiation*. As the name suggests, this process amounts, essentially, to applying facts and techniques concerning derivatives "in reverse."

4.2.2   Exercises

4.2.2.1 Part 1: Work as Force · Distance

The effort it takes to move an object is called **work**. Since it takes *twice* as much effort to move the object twice as far, or to move another object that is twice as heavy, we can see that the work done in moving an object is proportional to both the force applied and to the distance moved. The simplest way to express this fact is to define

$$\text{work} = \text{force} \cdot \text{distance.} \tag{4.14}$$

For example, to lift a weight of 20 lbs straight up takes 20 lbs of force. If the vertical distance is 3 ft, then

$$20 \text{ lbs} \cdot 3 \text{ ft} = 60 \text{ foot-pounds}$$

of work is done. (The *foot-pound* is one of the standard units for measuring work.)

The above Eq. (4.14) holds provided the applied force is *constant* over the entire distance in question. If not, then we treat work as an *accumulation function* for force. We can thereby make approximations to amounts of work done, using methods like those employed in this and the preceding section.

1. Suppose a tractor pulls a loaded wagon over a road whose steepness varies. If the first 150 ft of road are relatively level and the tractor has to exert only 200 lbs of force while the next 400 ft are inclined and the tractor has to exert 550 lbs of force, how much work does the tractor do altogether?
2. A motor on a large ship is lifting a 2000 lb anchor that is already out of the water at the end of a 30 foot chain. The chain weighs 40 lbs per foot. As the motor lifts the anchor, the part of the chain that is hanging gets shorter and shorter, thereby reducing the weight the motor must lift.

(a) What is the combined weight, call it $F(x)$, of anchor and hanging chain when the anchor has been lifted x feet above its initial position? Hint: when the anchor is x feet above its initial position, there are $30 - x$ feet of chain still hanging.

(b) Divide the 30-foot distance that the anchor must move into 3 equal intervals of 10 ft each. Estimate how much work the motor does lifting the anchor and chain over each 10-foot interval by multiplying the combined weight at the *bottom* of the interval by the 10-foot height. (That is: you'll be considering the weight of the chain at $x = 0$, $x = 10$, and $x = 20$.) What is your estimate for the total work

$$W = \int_0^{30} F(x)\, dx$$

done by the motor in raising the anchor and chain 30 ft?

(c) Repeat part (b) of this exercise, but this time, estimate work over each 10-foot interval by multiplying the combined weight at the *top* of the interval by the 10-foot height. (That is: you'll be considering the weight of the chain at $x = 10$, $x = 20$, and $x = 30$.) What is your new estimate for the total work W done in raising the anchor and chain 30 ft?

(d) Average your answers from parts (b) and (c) of this exercise, to obtain a better estimate of the amount of work done.

(e) Repeat all the steps of part (b), but this time use 6 equal intervals of 5 ft each. Is your new estimate of the work done larger or smaller than your estimate in part (b)? Which estimate is likely to be more accurate? On what do you base your judgment?

(f) If you ignore the weight of the chain entirely, what is your estimate of the work done? How much *extra* work do you therefore estimate the motor must do to raise the heavy chain along with the anchor?

4.2.2.2 Part 2: Distance as Velocity · Time

3. (a) Make a copy of Fig. 4.9. On your copy, over each of the ten intervals $[0.0.5]$, $[0.5, 1]$, ..., $[4, 4.5]$, $[4.5, 5]$, draw a rectangle whose height equals the value of $v(t)$ at the right endpoint of that interval.

(b) Use your above rectangles to approximate

$$d(T) = \int_0^T v(t)\, dt,$$

the distance traveled from time $t = 0$ to time $t = T$, for each of the times $t = 0, 0.5, \ldots, 4.5, 5$.

(c) Plot your points from part (b) of this exercise in the plane, and connect them with line segments, to get an approximation to the graph of $d(T)$ for $0 \leq T \leq 5$.

4.2.2.3 Part 3: Area Under a Graph

4. (a) Approximate the area

$$A = \int_3^7 \sqrt{1 + x^3}\,dx$$

under the graph of $f(x) = \sqrt{1 + x^3}$ on the interval $[3, 7]$, using a left endpoint approximation with four rectangles, all of equal baselength.

(b) Do you think your estimate is an overestimate or underestimate of the actual area A? Hint: you might want to sketch the function on this interval. Then consider the comments, following Example 4.2.3 above, about increase/decrease versus underestimates/overestimates.

(c) Repeat parts (a) and (b) of this exercise, but this time using a *right endpoint* approximation. (Everything else—the function, the interval, the number of rectangles—should remain the same as above.)

5. Repeat Example 4.2.3 above, but this time using a *right endpoint* approximation.

6. Repeat Example 4.2.3 above, but this time using a *midpoint* approximation. (The height of the rectangle over each of your ten subintervals should be equal to the value of the function $f(x) = \sqrt{x - 1}$ at the midpoint of that subinterval.)

4.3   More on Integration

In this section, we wish to put the ideas of the previous two sections on a somewhat more systematic footing.

4.3.1   Terminology and Notation

The process of evaluating an integral is called **integration**. Integration means "putting together." To understand why this name is appropriate, recall Sect. 4.1, where we estimated energy consumption over a long time interval by putting together a lot of energy contributions $p(t)\Delta t$ over a succession of short periods. Or consider Example 4.2.2 above, where we "assembled" approximate distances $v(t)\Delta t$, over shorter pieces of a trip, to estimate distance traveled over the entire five-hour journey. Or Example 4.2.3, where we used a sum of areas of rectangles to approximate the area under a curve.

The function being "accumulated," meaning integrated—for example, $p(t)$, or $v(t)$, or $\sqrt{x - 1}$—is called the **integrand**.

The numbers a and b in an integral

$$\int_a^b f(x)\,dx \tag{4.15}$$

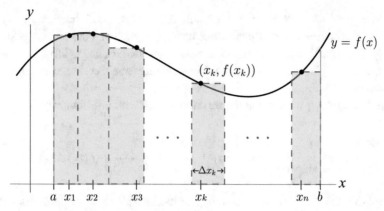

Fig. 4.11 Approximating the definite integral $\int_a^b f(x)\,dx$

are called the **limits of integration**, and the corresponding interval $[a, b]$ is called the **domain of integration**. The symbol dx in this integral is sometimes called a **differential**. Note that this differential has no numerical value *per se*. Think of it more as a sort of right bracket (where the symbol $\int_a^b$ is the corresponding left bracket).

As we've seen earlier, in a variety of circumstances, the integral (4.15) can be approximated by adding up areas of rectangles, whose bases lie on consecutive subintervals of $[a, b]$, and whose heights are given by values of f at "sampling points" x_k (Fig. 4.11).

In symbols,

$$\int_a^b f(x)\,dx \approx f(x_1)\Delta x_1 + f(x_2)\Delta x_2 + \cdots + f(x_n)\Delta x_n. \tag{4.16}$$

The sum on the right-hand side of (4.16) is called a **Riemann sum** for f on $[a, b]$. More specifically, we have the following.

Definition. Suppose the function f is defined for all x in the interval $[a, b]$. Then a **Riemann sum** for f on $[a, b]$ is an expression of the form

$$f(x_1)\Delta x_1 + f(x_2)\Delta x_2 + \cdots + f(x_n)\Delta x_n.$$

Here, the interval $[a, b]$ has been divided into n subintervals whose lengths are $\Delta x_1, \ldots, \Delta x_n$, respectively, and for each k from 1 to n, x_k is some point in the kth subinterval.

That is, Riemann sums are exactly the kinds of sums we've been using to approximate accumulation functions and definite integrals.

Intuitively, the more rectangles we use, and the narrower their baselengths, the closer these rectangles come to filling up the space under the graph of $f(x)$ precisely. So, at least intuitively, as the number n of rectangles increases *ad infinitum*, and as all of the baselengths Δx_k of these rectangles shrink to zero, our Riemann sum should converge *exactly* to the area A in question.

When this happens, we say that the function f is **(Riemann) integrable** on $[a, b]$, and we write

$$\int_a^b f(x)\,dx = \lim_{\substack{\Delta x_k \to 0 \\ \text{for all } k}} \left(f(x_1)\Delta x_1 + f(x_2)\Delta x_2 + \cdots + f(x_n)\Delta x_n \right), \qquad (4.17)$$

to indicate that the Riemann sums converge to the integral as all of the baselengths Δx_k shrink to zero.

There are functions f that *are not* Riemann integrable. But such functions arise infrequently in "real-world" contexts. Further, recall (see Remark 4.2.1) that any integrand f is tacitly assumed to be piecewise continuous on its interval $[a, b]$ of integration. And such functions are known to be Riemann integrable on such intervals.

Notice that neither the definition of Riemann sum, nor that of Riemann integrable function, specifies the choice of sampling points. The condition that the Riemann sums be close to a single number involves only the subintervals $\Delta x_1, \Delta x_2, \ldots, \Delta x_n$. This is important; it says *once the subintervals are small enough, it doesn't matter which sampling points x_k we choose—all of the Riemann sums will be close to the value of the integral*. (Of course, some will still be closer to the value of the integral than others.)

It is important to note that, from a mathematical point of view, a Riemann sum is just a number. It's the *context* that provides the meaning: Riemann sums for power demand that varies over time approximate total energy consumption; Riemann sums for a velocity that varies over time approximate total distance; Riemann sums for a force that varies over distance approximates total work done. And so on. Similarly, a definite integral is just a number, and it's the context that gives such an integral meaning.

4.3.2   Riemann Sums Using Technology

The Sage program RIEMANN, below, automates the left endpoint Riemann sum approximation of Example 4.2.3 above. The program has been written so that it may be easily modified, to accommodate different functions $f(x)$, different intervals $[a, b]$, different numbers n of rectangles, and different kinds of sampling points (other than just left endpoints).

Program RIEMANN

```
var ('x')            # our variable is called x
f(x) = sqrt(x-1)     # this is where you put your function
a = 1
b = 6
n = 10


Deltax = (b-a)/n     # baselength of the rectangles
                     # The following formula gives you a left endpoint sum

RS = sum(f(a+(k-1)*Deltax)*Deltax for k in [1..n])

print round(RS, 5)   #prints the output rounded to 5 decimal places
```

It's important to review the above program, to understand the purposes of the various lines. Such an understanding will make it easy to adapt the program to other situations.

The purpose of the first five lines is clear. The sixth line expresses the following fact: if the interval $[a, b]$ is to be divided into n subintervals of equal length Δx, then

$$\Delta x = \frac{b - a}{n}$$

Formula for baselength of rectangles, in a Riemann sum approximation where all rectangles have equal baselength

What does the line

$$RS = \text{sum}(f(a + (k - 1) * \text{Deltax}) * \text{Deltax for k in } [1..n]) \tag{4.18}$$

signify? Well, the crucial thing to note here is that the quantity $a + (k - 1)\Delta x$ represented in (4.18) *is* the left endpoint of the kth subinterval. To see this, think of it this way: to get to the left endpoint of the kth subinterval (assuming, again, that all subintervals have equal length Δx), you start at $x = a$, and perform $k - 1$ "jumps" of length Δx each (Fig. 4.12).

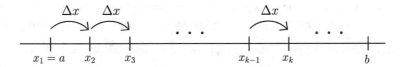

Fig. 4.12 The "jumps" required to get to the left endpoint x_k of the kth subinterval

So the line (4.18) is just saying: "To get your approximation RS, add up the numbers $f(x_k)\Delta x$, where x_k is the left endpoint of the kth subinterval, for $1 \le k \le n$." And this is exactly the sum we want for our left endpoint Riemann sum approximation.

We summarize:

> **Left endpoint Riemann sums: $x_k = a + (k-1)\Delta x$**

Formula for the sampling points x_k, in a left endpoint
Riemann sum approximation (with all intervals of equal length)

By modifying RIEMANN, you can calculate Riemann sums for other sampling points, other intervals, other numbers of rectangles, and other functions. For example, to sample at midpoints, you must start at $x = a$, and make $(k - 1) + 1/2 = k - 1/2$ jumps each of length Δx. To sample at right endpoints, you again start at $x = a$, but this time make k such jumps. So we have the following formulas:

> **Midpoint Riemann sums: $x_k = a + (k - 1/2)\Delta x$**
> **Right endpoint Riemann sums: $x_k = a + k\Delta x$**

Formulas for the sampling points x_k, in midpoint and right endpoint
Riemann sum approximations (with all intervals of equal length)

Of course, in a Riemann sum approximation, not all subintervals need to have the same length. We have already done a number of approximations where various different lengths were used in the same approximation. But equal lengths make for nicer formulas, and simpler code, and more systematic algorithms. We will use subintervals of equal length except where otherwise noted.

Example 4.3.1 Estimate the area

$$\int_1^6 \sqrt{x - 1}\, dx$$

of Example 4.2.3 above, using midpoint Riemann sums. Obtain a sequence of estimates, with more and more rectangles, until the fourth decimal place in your estimates stabilizes.

Solution. Because we want a midpoint approximation, we need to replace the line

```
RS = sum(f(a+(k-1)*Deltax)*Deltax for k in [1..n])
```

in the above program RIEMANN with the line

```
RS = sum(f(a+(k-1/2)*Deltax)*Deltax for k in [1..n])
```

(we replaced the (k-1) with (k-1/2)). Strictly speaking, we should also edit the comment line that refers to a left endpoint sum, though this will not affect out output.) Let's also start with n=100 rectangles.

We run this modified version of RIEMANN, and get an output of 7.45422.

If we then repeat with n=200 rectangles, our estimate is 7.45379. Continuing with n=500, n=1000, n=2500, and n=5000, we get the estimates

$$7.45362, \quad 7.45358, \quad 7.45357, \quad 7.45356,$$

respectively. In the last three of these, the fourth decimal place equals 5. So, to four decimal places, the area in question appears to be equal to 7.4535.

4.3.3 Exercises

4.3.3.1 Part 1: Using RIEMANN

For the exercises in this part, do not try to write down, or compute, the requested Riemann sums by hand. Simply use the above program RIEMANN, suitably modified.

1. Calculate left endpoint Riemann sums for the function $\sqrt{1 + x^3}$ on the interval $[3, 7]$ using 10, 100, 1,000, and 10,000 equally-spaced subintervals. (You may need to be a bit patient when n is as large as 10,000.) Note that these Riemann sums approximations seem to be approaching a limit—that is, zeroing in on some particular number—as the number of subintervals gets larger and larger. What does that limit seem to be, rounded to the nearest hundredth?
2. Repeat Exercise 1 above for the same function, but this time, on the interval $[1, 3]$.
3. Repeat Exercise 2 above, but this time, use *right endpoint* Riemann sums.
4. (a) Repeat Exercise 2 above, but this time, use *midpoint* Riemann sums.
   (b) Comment on the relative "efficiency" of midpoint Riemann sums, versus left and right endpoint Riemann sums (at least for the function $\sqrt{1 + x^3}$ on the interval $[1, 3]$). (A more efficient procedure is one that will "zero in" on a particular value faster than a less efficient one.)
5. Calculate left endpoint Riemann sums for the function

$$f(x) = \cos(x^2) \quad \text{on the interval } [0, 4],$$

   using 100, 1000, and 10000 equally-spaced subintervals.
6. Use right endpoint Riemann sums, with 10, 100, and 1000 equally-spaced subintervals, to estimate the integral

$$\int_2^3 \frac{\cos(x)}{1 + x^2} \, dx.$$

The Riemann sums are all negative; why? (A suggestion: sketch the graph of f, preferably by computer. What does that tell you about the signs of the terms in a Riemann sum for f?)

7. (a) Calculate midpoint Riemann sums for the function

$$H(z) = z^3 \quad \text{on the interval } [-2, 2],$$

using 10, 100, and 1000 equally-spaced subintervals. The Riemann sums are all zero; why?

   (b) Repeat part (a) using *left endpoint* Riemann sums. Are the results still zero? Can you explain the difference, if any, between these two results?

8. Using RIEMANN, obtain a sequence of estimates for each of the following integrals. Continue until the first three decimal places stabilize in your estimates.

(a) $\int_0^1 x^3 \, dx$ (b) $\int_0^3 x^3 \, dx$ (c) $\int_0^\pi x \sin(x) \, dx$ (d) $\int_0^1 \frac{1}{1+x^3} \, dx$ (e) $\int_1^2 e^{-x^2} \, dx$

(Remember that, in Sage, π is entered as `pi`.)

9. What is the area under the curve $y = e^{-x^2}$ over the interval $[0, 1]$? Give an estimate that is accurate to three decimal places. Sketch the curve and shade the area.

4.3.3.2 Part 3: Making Approximations Using Riemann Sums

10. **Waste production.** A colony of living yeast cells in a vat of fermenting grape juice produces waste products—mainly alcohol and carbon dioxide—as it consumes the sugar in the grape juice. It is reasonable to expect that another yeast colony, twice as large as this one, would produce twice as much waste over the same time period. Moreover, since waste accumulates over time, if we double the time period we would expect our colony to produce twice as much waste.

These observations suggest that waste production is proportional to both the size of the colony and the amount of time that passes. If P is the size of the colony, in grams, and Δt is a short time interval, then we can express waste production W as a function of P and Δt:

$$\Delta W \approx k \cdot P \cdot \Delta t \text{ grams.}$$

If Δt is measured in hours, then the multiplier k has to be measured in units of grams of waste per hour per gram of yeast.

The preceding formula is useful only over a time interval Δt in which the population size P does not vary significantly. If the time interval is large, and the population size can be expressed as a function $P(t)$ of the time t, then we can estimate waste production by breaking up the whole time interval into a succession of smaller intervals $\Delta t_1, \Delta t_2,$ ..., Δt_n and forming a Riemann sum

$$k \, P(t_1) \, \Delta t_1 + \cdots + k \, P(t_n) \, \Delta t_n \approx W \text{ grams.}$$

The time t_k must lie within the time interval Δt_k. For Δt_k small enough, $P(t_k)$ should be a good approximation to the population size $P(t)$ throughout that time interval. Suppose the colony starts with 300 g of yeast (i.e., at time $t = 0$ h) and it grows exponentially according to the formula

$$P(t) = 300\,e^{0.2t}\,.$$

If the waste production constant k is 0.1 g per hour per gram of yeast, estimate how much waste is produced in the first four hours. Use a Riemann sum with four hour-long time intervals and measure the population size of the yeast in the middle of each interval—that is, "on the half-hour." (So $t_1 = 0.5$, $t_2 = 1.5$, and so on; and all Δt_k's are equal to 1.)

11. **The Colorado flood.** In September 2013, a cold front over Colorado collided with monsoonal air flowing from the south. The result was heavy rain of unusually long duration, causing catastrophic flooding along the eastern foothills of the Colorado Rocky Mountains. Boulder County was severely impacted, with several deaths; complete loss of several hundred homes, with many more extensively damaged; and numerous stretches of road completely swept away.

    The graph below depicts Boulder Creek flow during the September 2013 Colorado flood, as measured by the U.S. Geological Survey (USGS). The graph uses a logarithmic scale; that is, the horizontal lines above 100 represent 200, 300, 400, 500, 600, 700, 800, 900, 1000, then 2000, 3000, etc. (The term "discharge" on the vertical axis is synonymous with "flow rate.")

    The small asterisks represent certain redundant measurements that were used for calibration. You needn't worry about these.

    (a) What was the approximate flow rate at noon on Thursday, September 12, 2013?

    (b) Using only your answer to part (a) of this exercise, approximate the quantity of water (in cubic feet) that flowed through Boulder Creek at this station during the day (24-hour period) of Thursday, September 12, 2013. Be careful: the units on the vertical axis are cubic feet per *second*.

    (c) What is the 27-year median flow rate (the "median daily statistic") for September 12? (The median is a kind of average, so the median daily statistic is a kind of measure of the average flow rate for the day in question, averaged over a certain set of 27 years.) Approximate the median, for these 27 years, of the *total amount* of water that flowed through Boulder Creek at this station on the 12th of September.

    (d) Make a copy of the graph below. Then on this copy, draw rectangles, with $\Delta t = 12$ h for each rectangle, representing a left-endpoint Riemann sum approximation to the the total quantity of water that flowed through Boulder Creek at this station, from the beginning of Wednesday, September 11 through the end of Tuesday, September 17.

(e) Use your rectangles from the previous part of this exercise to estimate the total water
flow through Boulder Creek over the seven-day period in question. You needn't
worry too much about interpolating the logarithmic scale. For example, you could
simply round each flow rate measurement to the nearest horizontal grid line.
What are your units for this estimate?

(f) Use the median daily statistic data, and a Riemann sum with $\Delta t = 24$ h, to approximate the median quantity of water that flowed through the creek over the course of
these same dates, for the 27-year period considered. (The orange triangles are not
very precise, so make some educated guesses.)

(For more flow data, see http://waterdata.usgs.gov/nwis/uv?06730200)

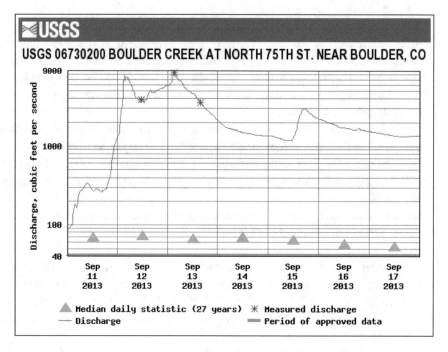

4.4   The Geometry of Definite Integrals

Some integrals can be evaluated *exactly*, using simple geometric considerations. Let's consider a couple of integrals that can be determined in this way—and one that can't, to set the
stage for the next section.

Example 4.4.1 Evaluate each of the following integrals using ideas from basic geometry, or explain why this is not possible. (If it helps, sketch the indicated function over the given interval.)

$$\text{(a) } \int_0^5 4\,dx \qquad\qquad \text{(b) } \int_0^5 4x\,dx \qquad\qquad \text{(c) } \int_0^5 4x^2\,dx$$

Solution. (a) This integral represents the area under the graph of $f(x) = 4$, a constant function of height 4, over the interval $[0, 5]$. The region in question is just a rectangle of baselength 5 (=the length of $[0, 5]$), and height 4 (=the height of $f(x)$). So $\int_0^5 4\,dx = 5 \cdot 4 = 20$.

(b) This integral represents the area under the graph of $f(x) = 4x$, a linear function, over the interval $[0, 5]$. The region in question is just a triangle of baselength 5 (= the length of $[0, 5]$) and height $4 \cdot 5 = 20$ (= the height of $f(x)$ at $x = 5$). So $\int_0^5 4x\,dx = \frac{1}{2} \cdot 5 \cdot 20 = 50$.

(c) This integral represents the area under the graph of $f(x) = 4x^2$, a parabola, over the interval $[0, 5]$. The region in question has a "curved top;" this region doesn't look like any familiar shape from high school geometry. For now, we have no immediate way of determining its area exactly.

Note that, for part (c) of the above example, we could certainly *estimate* the area in question, to any desired accuracy, using Riemann sums. For example, applying the above program RIEMANN to a *midpoint* approximation with $f(x) = 4x^2$, $a = 0$, $b = 5$, and $n = 5,000$ gives us the approximation $RS = 166.66666$ to this integral.

In the next section, we'll see how to evaluate this integral *exactly*, and we'll find that $\int_0^5 4x^2\,dx = 500/3$, which, rounded to five decimal places, equals 166.66667.

4.4.1   The Integral of a (sometimes) Negative Function

Up to this point, we have been dealing with a function $y = f(x)$ that is never negative on the interval $[a, b]$. Its graph therefore lies entirely above the x-axis. What happens if f *does* take on negative values on this interval? To answer, we consider the graph below (Fig. 4.13).

The figure illustrates the fact that, if $f(x) < 0$ on some portion of the interval $[a, b]$, then some of the summands $f(x_k)\Delta x_k$ in the Riemann sum of (4.16) will, typically, be negative. And these negative contributions will impact the limit on the right-hand side of (4.17).

So let's imagine a series of Riemann sum rectangles spanning $[a, b]$, in Fig. 4.13. And let's think about which rectangles lie above the x-axis, and therefore contribute a *positive* amount to the Riemann sum, and which rectangles lie below the x-axis, and therefore contribute a *negative* amount to the Riemann sum. If we then consider what happens as all of these rectangles become very narrow, we are led to the following conclusion.

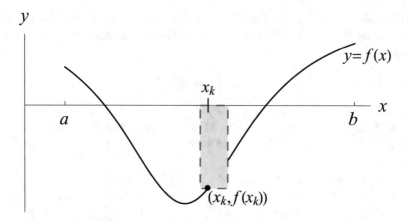

Fig. 4.13 A function that is sometimes positive and sometimes negative on an interval $[a, b]$, and a representative rectangle for a Riemann sum approximating $\int_a^b f(x)\,dx$

$$\int_a^b f(x)\,dx \text{ is equal to the } \textit{signed area} \text{ between}$$

the graph of $y = f(x)$ and the x-axis, on the interval $[a, b]$,
meaning the sum of areas of regions above the x-axis,
minus the sum of areas of regions below the x-axis.

Geometric interpretation of the integral, for functions
that may sometimes be negative on an interval

Example 4.4.2 (a) Consider the graph of $y = x$ over the interval $[-2, 4]$.

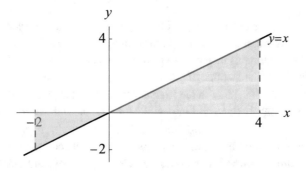

The upper region (shaded in blue) is a triangle whose area is $\frac{1}{2} \cdot 4 \cdot 4 = 8$. The lower region (shaded in red) is another triangle; its area $\frac{1}{2} \cdot 2 \cdot 2 = 2$. Thus, the *signed area* between the graph of $y = f(x)$ and the x-axis, on the interval $[-2, 4]$, is $8 - 2 = 6$. It follows that

$$\int_{-2}^{4} x \, dx = 6.$$

You should confirm that Riemann sums for $f(x) = x$ over the interval $[-2, 4]$ converge to the value 6. See the Exercises below. (We'll evaluate this integral in another way in the next section.)

(b) $\int_{-\pi}^{\pi} \sin(x) \, dx = 0$, since the areas of the two "lobes"—one above the x-axis and one below—cancel.

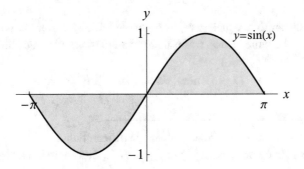

Again, you should confirm this result using Riemann sums. (And we'll verify it using another approach in the next section.)

4.4.2   Integration Rules

Just as there are rules that tell us how to find the derivative of various combinations of functions, there are other rules that tell us how to find the integral. Here are two rules that are exactly analogous to differentiation rules:

$$\int_{a}^{b} f(x) + g(x) \, dx = \int_{a}^{b} f(x) \, dx + \int_{a}^{b} g(x) \, dx \qquad \textbf{(sum rule for integrals)};$$

$$\int_{a}^{b} c f(x) \, dx = c \int_{a}^{b} f(x) \, dx \qquad \textbf{(constant multiple rule for integrals)}.$$

The first rule says that the signed area of a sum of two functions is the sum of the signed areas of the original two functions. The second rule says that, if you rescale a function by a factor of c in the vertical direction, then you rescale its signed area by that same factor. (And also that, if you multiply a function by a negative number c, then the new function's signed area has the opposite sign from that of the original function.)

Example 4.4.3 Use geometry, together with the sum and constant multiple rules, to compute

$$\int_{-\pi}^{\pi} (7 - 3\sin(x))\, dx.$$

Solution. By the above two rules,

$$\int_{-\pi}^{\pi} (7 - 3\sin(x))\, dx = \int_{-\pi}^{\pi} 7\, dx + \int_{-\pi}^{\pi} (-3\sin(x))\, dx = \int_{-\pi}^{\pi} 7\, dx - 3\int_{-\pi}^{\pi} \sin(x)\, dx.$$

The integral on the far right equals zero, by Example 4.4.2(b) above. Moreover, $\int_{-\pi}^{\pi} 7\, dx = 2\pi \cdot 7 = 14\pi$, by the same idea as was used in Example 4.4.1(a) above. So

$$\int_{-\pi}^{\pi} (7 - 3\sin(x))\, dx = 14\pi - 3 \cdot 0 = 14\pi.$$

Here are two rules that have no direct analogue in differentiation.

Comparison rule for integrals: If $f(x) \leq g(x)$ for every x in the interval $[a, b]$, then

$$\int_a^b f(x)\, dx \leq \int_a^b g(x)\, dx.$$

Juxtaposition rule for integrals: If c is a point somewhere in the interval $[a, b]$, then

$$\int_a^b f(x)\, dx = \int_a^c f(x)\, dx + \int_c^b f(x)\, dx.$$

(The sum rule tells us how integrals "add vertically;" the juxtaposition rule tells us how they "add horizontally.")

If you visualize an integral as an area, you can see why the above two rules are true.

Our last rule allows us to do things like "integrate from 7 to 2," or more generally to integrate from a larger number to a smaller one. It says

$$\int_b^a f(x)\, dx = -\int_a^b f(x)\, dx \qquad \textbf{(reversal rule for integrals)}.$$

So for instance, $\int_5^0 4x\, dx = -50$, by the reversal rule and by Example 4.4.1(b) above.

One way to think of the reversal rule is as follows: suppose $a \leq b$. If we imagine that integrating f from a to b is like traveling along $[a, b]$ from left to right, and painting the regions bounded by the graph of f, then integrating from b to a is like running that process in reverse, and "unpainting" those areas.

Still, from the point of view of mathematics and its applications, it might seem unnecessary to define integrals like $\int_5^0 4x\, dx$. As we will see, though, integrals from larger to smaller numbers *do* arise—for example, in integration by substitution, which we will consider in a later section.

4.4.3 Exercises

For these exercises, you should use geometry (**not** Riemann sums), together with integration rules where necessary, to evaluate the integrals in question.

1. Determine the values of the following integrals. Explain your reasoning.

(a) $\displaystyle\int_{2}^{15} 3\,dx$ (b) $\displaystyle\int_{2}^{15} 3x\,dx$ (c) $\displaystyle\int_{15}^{2} 3x\,dx$ (d) $\displaystyle\int_{-\pi}^{\pi} (4\sin(x)+3)\,dx$

(e) $\displaystyle\int_{-\pi}^{\pi} (\sin(x)+3x)\,dx$ (f) $\displaystyle\int_{\pi}^{-\pi} (\sin(x)+3x)\,dx$ (g) $\displaystyle\int_{-4}^{9} (4-x)\,dx$

2. (a) Sketch the graph of

$$g(x) = \begin{cases} 7 & \text{if } 1 \le x < 5, \\ -3 & \text{if } 5 \le x \le 10. \end{cases}$$

(b) Determine $\displaystyle\int_{1}^{7} g(x)\,dx,\ \int_{7}^{10} g(x)\,dx,$ and $\displaystyle\int_{1}^{10} g(x)\,dx.$

3. Repeat Exercise 2 above for

$$g(x) = \begin{cases} x-2 & \text{if } 1 \le x < 6, \\ 4 & \text{if } 6 \le x \le 10. \end{cases}$$

4. Below is a picture of the graph of a function $y = f(x)$. Use geometry and properties of integrals to evaluate the indicated definite integrals.

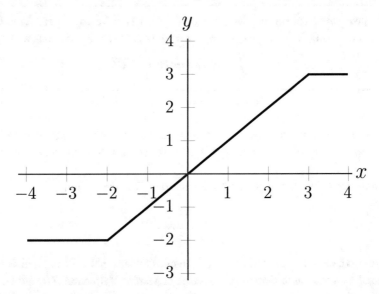

(a) $\displaystyle\int_{-3}^{4} f(x)\,dx$

(b) $\displaystyle\int_{0}^{3} (4f(x) - 3)\,dx$

(c) $\displaystyle\int_{-4}^{1} f(x)\,dx + \int_{1}^{-4} f(x)\,dx$

5. (a) Sketch, by hand or computer, the graphs of $y = \cos(x)$ and $y = 5 + \cos(x)$ over the interval $[0, 4\pi]$.

   (b) Find $\displaystyle\int_{0}^{4\pi} \cos(x)\,dx$ by interpreting the integral as a signed area.

   (c) Find $\displaystyle\int_{0}^{4\pi} 5 + \cos(x)\,dx$. Why does $\displaystyle\int_{0}^{4\pi} 5\,dx$ have the same value?

4.5   The Fundamental Theorem of Calculus

4.5.1   Statement and Discussion

The following central result, known as the Fundamental Theorem of Calculus, tell us that we can **integrate**—that is, evaluate a definite integral—by **antidifferentiating**—that is, by performing a process that is "reverse" to differentiation.

> **Suppose p is a continuous function on $[a, b]$. Let E be any *antiderivative* of p on (a, b), meaning $E'(t) = p(t)$ for $a < t < b$. If E is also continuous on $[a, b]$, then**
> $$\int_{a}^{b} p(t)\,dt = E(b) - E(a).$$

The Fundamental Theorem of Calculus

Shortly, we'll outline a proof of the Fundamental Theorem of Calculus. But first we provide an example, and a few comments, to highlight the theorem's remarkable power.

Example 4.5.1   Use the Fundamental Theorem of Calculus to find

$$\int_{0}^{\pi/2} \cos(t)\,dt. \tag{4.19}$$

Solution. The Fundamental Theorem of Calculus tells us that we can evaluate this integral if we can first find an *antiderivative* of $\cos(t)$, meaning a function $E(t)$ such that $E'(t) = \cos(t)$.

At the moment, the only method we have of antidifferentiating (finding antiderivatives) is to try and work backwards from known derivative formulas and facts. Specifically: to find an antiderivative of $p(t) = \cos(t)$, we recall, from our studies of the derivative, that the cosine and sine functions are closely linked to each other through differentiation. In particular, we recall that

$$\frac{d}{dt}\big[\sin(t)\big] = \cos(t).$$

Therefore, if we choose $E(t) = \sin(t)$, then we do in fact have

$$E'(t) = \frac{d}{dt}\big[\sin(t)\big] = \cos(t) = p(t).$$

So by The Fundamental Theorem of Calculus,

$$\int_0^{\pi/2} \cos(t)\,dt = E(\pi/2) - E(0) = \sin(\pi/2) - \sin(0) = 1 - 0 = 1.$$

We make a few observations regarding the above theorem and example.

1. There's no simple geometric argument (like those of Example 4.4.2 above) that applies to evaluate the integral (4.19) directly. Of course, we could have computed this integral, at least approximately, using Riemann sums. But the Fundamental Theorem of Calculus provides us with a method that is more straightforward, immediate, and exact.
2. Of course, the Fundamental Theorem of Calculus does require that we find an *antiderivative* of the integrand. And this can present problems. See, for instance, Example 4.5.3(e) below. For this reason, the Riemann sum method remains relevant—in fact, critical— in a great variety of circumstances.

   Still, when it does work, the Fundamental Theorem of Calculus is the method of choice, because of its simplicity and exactness.
3. The requirements on E and p, in the statement of the Fundamental Theorem of Calculus, may be relaxed somewhat. In particular, the function p need only be *piecewise continuous*, rather than strictly continuous, on $[a, b]$. (See Remark 4.2.1, above, for the definition of piecewise continuity.) For such functions p, the conclusion of the Fundamental Theorem remains true even though, typically, E' will fail to exist at those points where p is discontinuous.

   All of this is exemplified by Exercise 14 below. In that exercise, we show that the Fundamental Theorem of Calculus provides the correct result for the energy function E and the power function p of Example 4.1.4 above, even though the function p there is only piecewise continuous on the interval in question.
4. The technical details of the Fundamental Theorem, and of the conditions under which its conclusion holds, are of course very important. Of perhaps greater importance, though, is to appreciate the *gist* of the theorem, which is, again: we can **integrate** by **antidifferentiating**.

5. We presented the Fundamental Theorem in terms of a variable t, and functions p and E, to evoke our discussions from Sect. 4.1 of energy E and power p, as functions of time t. In many other references, this theorem is presented as follows:

Suppose f is a continuous function on $[a, b]$. If F is continuous on $[a, b]$ and is an antiderivative of f on (a, b), then

$$\int_a^b f(x)\, dx = F(b) - F(a).$$

The Fundamental Theorem of Calculus

This is nothing more than our original presentation, with t replaced by x, p by f, and E by F. Still, the use of x's, f's, and F's has become ubiquitous, so it's worthwhile to give both presentations of the theorem.

4.5.2 Sketch of a Proof of the Fundamental Theorem of Calculus

We've devoted a considerable portion of this text to the theory of initial value problems. We now use certain aspects of this theory to provide a (somewhat informal, but readily "formalizeable") proof of the Fundamental Theorem of Calculus.

The proof amounts, essentially, to four facts:

Fact 1. The integral

$$I(T) = \int_a^T p(t)\, dt \qquad (a \le T \le b) \tag{4.20}$$

is continuous on $[a, b]$ and solves the initial value problem

$$I' = p \text{ on } (a, b); \qquad I(a) = 0. \tag{4.21}$$

Fact 2. If E is a continuous function on $[a, b]$ such that $E'(t) = p(t)$ for $a < t < b$, then the function $G(t)$ defined by

$$G(t) = E(t) - E(a) \tag{4.22}$$

is **also** a solution to (4.21).

Fact 3. By the theory of initial value problems there is, in fact, only *one* continuous function satisfying (4.21). **So the functions I and G must be the same**—that is, we conclude that $I(t) = G(t)$ for $a \le T \le b$.

Fact 4. In particular, $I(b) = G(b)$, meaning the left-hand sides of (4.20) and (4.22) are equal at $T = b$. But then so are the right-hand sides, so that

$$\int_a^b p(t)\,dt = E(b) - E(a), \tag{4.23}$$

where, again, E is a function with $E' = p$ on (a, b). But (4.23) **is** the conclusion of the Fundamental Theorem of Calculus. So **Fact 4** completes our (informal) proof.

Actually, the completion of this proof still requires that we verify **Fact 1** through **Fact 4**. We do so in reverse order. **Fact 4** follows immediately from **Fact 3**. **Fact 3** itself is a well-known result; we omit the proof. **Fact 2** is straightforward: given that E is an antiderivative of p on (a, b), we have

$$G'(t) = \frac{d}{dt}[E(t) - E(a)] = E'(t) - 0 = p(t)$$

on (a, b), since $E' = p$ there, and because $E(a)$ is constant with respect to t. So G satisfies the differential equation in (4.21). It also satisfies the initial condition there, since

$$G(a) = E(a) - E(a) = 0.$$

Next, we come to **Fact 1.** To see that $I(t)$ satisfies the specified initial condition $I(a) = 0$, we simply observe that, by (4.20), we have

$$E(a) = \int_a^a p(t)\,dt = 0,$$

since the length of the interval $[a, a]$ is zero, and therefore the area of the (very, very thin) vertical strip lying between this (very, very narrow) interval and the graph of $p(t)$ is also zero.

Finally, we consider the differential equation in (4.21). To see that the integral I defined by (4.20) satisfies this equation, we recall that I is an accumulation function for p. Therefore,

$$\Delta I \approx p(t)\Delta t$$

over any interval of sufficiently small length Δt, with t being any point in that interval. (See Eq. (4.11) above.) Dividing both sides by Δt gives

$$p(t) \approx \frac{\Delta I}{\Delta t}; \tag{4.24}$$

letting $\Delta t \to 0$ then gives

$$p(t) = \lim_{\Delta t \to 0} \frac{\Delta I}{\Delta t} = \frac{dI}{dt} = I'(t). \tag{4.25}$$

(The "$\approx$" becomes "$=$" in the limit because, the smaller Δt is, the less I should vary on the interval in question, so the better the approximation should be.) So the integral I does indeed satisfy the differential equation $I' = p$ specified in (4.21).

Thus we see that the integral I defined by (4.20) satisfies both the differential equation and the initial condition stipulated in the initial value problem (4.21), and is therefore a solution to that initial value problem. So we have completed our verification of **Fact 4** and, in turn, of the Fundamental Theorem of Calculus itself.

Again, our proof of this theorem is not entirely rigorous, but the ideas behind it have rigorous underpinnings.

4.5.3   More Examples and Observations

Example 4.5.2   Use The Fundamental Theorem of Calculus to evaluate each of the integrals in Example 4.4.1.

Solution. The integrals in question are

$$\int_0^5 4\,dx, \qquad \int_0^5 4x\,dx, \qquad \int_0^5 4x^2\,dx.$$

To evaluate any one of these integrals using The Fundamental Theorem of Calculus, we need to first find an antiderivative F of each integrand f in question.

For now, the only method we have of antidifferentiating is to try and work backwards from known derivative formulas and facts. (Later, we'll consider other techniques, though even these rely, ultimately, on "reversing" differentiation thought processes.) For example: to find an antiderivative of $f(x) = 4$, we ask: what kind of function has a constant derivative? To answer, we recall that the derivative measures slope, and the functions with constant slopes are the linear functions. In particular, we see that $F(x) = 4x$ is an antiderivative of $f(x) = 4$, since $F'(x) = d[4x]/dx = 4 = f(x)$. So, by The Fundamental Theorem of Calculus,

$$\int_0^5 4\,dx = F(5) - F(0) = 4 \cdot 5 - 4 \cdot 0 = 20,$$

as we found previously.

To integrate $f(x) = 4x$, we seek a function whose derivative is f. What kind of function "differentiates down" to $4x$? Well, we recall from many differentiation examples that the derivative of x^2 is $2x$. Adjusting by a constant factor, we find that the derivative of $F(x) = 2x^2$ is $f(x) = 4x$. So this is the $F(x)$ we want. Then by The Fundamental Theorem of Calculus,

$$\int_0^5 4x\,dx = F(5) - F(0) = 2 \cdot 5^2 - 2 \cdot 0^2 = 50,$$

also agreeing with previous results.

To integrate $f(x) = 4x^2$ we "work backwards" in a similar fashion to find that $F(x) = \dfrac{4}{3}x^3$ has the desired derivative $f(x)$:

$$F'(x) = \frac{d}{dx}\left[\frac{4}{3}x^3\right] = \frac{4}{3} \cdot \frac{d}{dx}[x^3] = \frac{4}{3} \cdot 3x^2 = 4x^2 = f(x).$$

So by The Fundamental Theorem of Calculus,

$$\int_0^5 4x^2 \, dx = F(5) - F(0) = \frac{4}{3} \cdot 5^3 - \frac{4}{3} \cdot 0^3 = \frac{500}{3},$$

a result that we were *not* able to obtain previously, without resorting to Riemann sums.

Before proceeding further, we introduce some notation that will allow us to write our answers more immediately.

Definition 4.5.1 If F is a function defined at both $x = a$ and $x = b$, then we write

$$F(x)\Big|_a^b$$

to denote the difference $F(b) - F(a)$.

The Fundamental Theorem of Calculus now tells us that, if F is an antiderivative of f, then

$$\int_a^b f(x) \, dx = F(x)\Big|_a^b.$$

This new notation allows us to write out solutions to integrals compactly. For example, we can now write

$$\int_0^5 4x^2 \, dx = \frac{4}{3}x^3\Big|_0^5 = \frac{4}{3} \cdot 5^3 - \frac{4}{3} \cdot 0^3 = \frac{500}{3},$$

without ever having to write things like "let $F(x) = 4x^3/3$" explicitly.

Example 4.5.3 Evaluate the following definite integrals, using the Fundamental Theorem of Calculus.

(a) $\int_0^1 e^x \, dx$ (b) $\int_0^1 e^{7x} \, dx$ (c) $\int_0^{\pi/4} \cos(2x) \, dx$ (d) $\int_1^6 \sqrt{x-1} \, dx$ (e) $\int_0^1 e^{-x^2/2} \, dx$

Solution. (a)

$$\int_0^1 e^x \, dx = e^x\Big|_0^1 = e^1 - e^0 = 1.71828.$$

Here, we've used the fact that e^x is an antiderivative of e^x, since $d[e^x]/dx = e^x$.

   (b) An antiderivative of e^{7x} is $e^{7x}/7$. This may not be immediately obvious; antidifferentiation sometimes requires some "guessing and checking." That is: we might guess, based for instance on part (a) of this example, that an antiderivative of e^{7x} is e^{7x}. But then we'd check that $d[e^{7x}]/dx = 7e^{7x}$. So we have an extra factor of 7; we can compensate for this by dividing our original guess by 7. This works, since

$$\frac{d}{dx}\left[\frac{e^{7x}}{7}\right] = \frac{1}{7} \cdot \frac{d}{dx}[e^{7x}] = \frac{1}{7} \cdot 7e^{7x} = e^{7x}.$$

So

$$\int_0^1 e^{7x}\, dx = \left.\frac{e^{7x}}{7}\right|_0^1 = \frac{e^{7\cdot1}}{7} - \frac{e^{7\cdot0}}{7} = \frac{e^7 - 1}{7} = 156.51902.$$

(c) $\displaystyle\int_0^{\pi/4} \cos(2x)\, dx = \left.\frac{\sin(2x)}{2}\right|_0^{\pi/4} = \frac{\sin(2\cdot\pi/4) - \sin(2\cdot0/4)}{2} = \frac{\sin(\pi/2) - \sin(0)}{2}$

$= \dfrac{1-0}{2} = \dfrac{1}{2}.$

(d) Since differentiation of a power decreases the exponent by one, we would expect antidifferentiation to have the reverse effect. So we might guess that an antiderivative of $\sqrt{x-1} = (x-1)^{1/2}$ is $(x-1)^{3/2}$. But

$$\frac{d}{dx}\left[(x-1)^{3/2}\right] = \frac{3}{2}(x-1)^{1/2}\frac{d}{dx}[(x-1)] = \frac{3}{2}(x-1)^{1/2},$$

so our guess is off by a factor of 3/2. Our new guess of $2(x-1)^{3/2}/3$ then works as an antiderivative of $\sqrt{x-1}$, as you should check. So

$$\int_1^6 \sqrt{x-1}\, dx = \left.\frac{2}{3}(x-1)^{3/2}\right|_1^6 = \frac{2}{3}\left((6-1)^{3/2} - (1-1)^{3/2}\right) = \frac{2\cdot5^{3/2}}{3} = \frac{2\sqrt{125}}{3} = 7.45356.$$

Compare this with the result of Example 4.2.3.

(e) To evaluate $\int_0^1 e^{-x^2/2}\, dx$ using The Fundamental Theorem of Calculus, we'd need to find an antiderivative of $e^{-x^2/2}$. But as it turns out, there *is* no such antiderivative that can be written in closed form.

What this example illustrates is that the The Fundamental Theorem of Calculus doesn't always work. It also illustrates how philosophically different antidifferentiation is from differentiation. Specifically, if we can write something down in terms of familiar functions and familiar mathematical operations, then we can differentiate it, in terms of familiar functions and familiar mathematical operations. With antidifferentiation, we're not so lucky, as the function $f(x) = e^{-x^2/2}$ demonstrates.

Fortunately, we always have Riemann sums as a recourse. And in this case we're *very* fortunate, because the integral $\int_0^1 e^{-x^2/2}\, dx$, and similar ones, show up all over the place in probability and statistics. (The function $f(x) = e^{-x^2/2}\, dx$ represents an example of a *normal* curve.) We'd be quite stuck if we weren't able to evaluate (or at least approximate, to whatever degree of accuracy we need) such integrals at all.

An integral is an area under a curve (or, more generally, a signed area), as described in the previous section. But again, we are not interested in integrals simply for the sake of studying areas of geometric objects. We study integrals largely because of their relevance to *accumulation functions*. Consider, for example, how we expressed the energy consumption of a town over a 24-hour period. The basic relation

$$\text{energy} = \text{power} \cdot \text{elapsed time}$$

could not be used directly, because power demand varies. Indirectly, though, we found that we could use this relation to build a Riemann sum for power demand p over time. This gave us an *approximation*:

$$\text{energy} \approx p(t_1)\Delta t_1 + p(t_2)\Delta t_2 + \cdots + p(t_n)\Delta t_n \quad \text{megawatt-hours.}$$

As these sums are refined (that is, more and more, narrower and narrower, rectangles are used), they converge to the true level of energy consumption, which we denote by a definite integral of the power function p. That is,

$$\text{energy} = \int_0^{24} p(t)\, dt \quad \text{megawatt-hours.} \tag{4.26}$$

In other words, *energy is the integral of power over time.*

In Example 4.2.2 we asked how far a car would travel in 5 h if we knew its velocity was $v(t)$ miles per hour at time t. We estimated that distance using Riemann sums, which converge to the integral

$$\int_0^5 v(t)\, dt \quad \text{miles.} \tag{4.27}$$

In other words, *distance is the integral of velocity over time.*

By similar reasoning, *work is the integral of force over distance.* (See the Exercises in Sect. 4.2.2.) In general, whenever F is an accumulation function for a function f on an interval $[a, b]$, we have

$$\Delta F \text{ on } [a, b] = \int_a^b f(x)\, dx. \tag{4.28}$$

The energy integral (4.26) has the same units as the Riemann sums that approximate it. Its units are the product of the megawatts used to measure p and the hours used to measure t. The units for the distance integral (4.27) are the product of the miles per hour used to measure velocity and the hours used to measure time. In general, the units for the integral in (4.28) are the product of the units for f and the units for x.

All of the above contexts, together with the definition itself of the the definite integral, suggest an intuitive interpretation of integration. Specifically: recall that a Riemann sum for a function f on an interval $[a, b]$ is a (finite) sum of terms of size $f(x_k)\Delta x_k$. These Riemann sums tend to the integral

$$\int_a^b f(x)\,dx$$

as the "widths" Δx_k of these terms tend to zero. On the other hand, as the Δx_k's tend to zero, there are more and more of these terms, and they become narrower and narrower. So we can *imagine* that this integral equals "the *continuous* sum, over all x between a and b, of terms of *infinitesimal* width dx and overall size $f(x)\,dx$."

It's difficult to give precise mathematical meaning to the notions of "continuous sum" and "infinitesimal width." The mathematician Abraham Robinson was able to do so, rigorously, in the 1960's, but his work (on "nonstandard analysis") is far afield from where we are now. We'll be content with the above interpretation as a heuristic way of seeing things.

With this interpretation in mind, and with the Fundamental Theorem of Calculus as a critical tool, we can evaluate various accumulation functions, provided we have a formula for the integrand, and for the antiderivative of this integrand.

Example 4.5.4 Find the amount of work done if a force of $F(x) = 2 + \cos(x)$ pounds is applied from $x = 0$ to $x = 6\,\text{ft}$.

Solution. We are "adding up" infinitesimal "work elements" of size $F(x)\,dx = (2 + \cos(x))\,dx$, from $x = 0$ to $x = 6$. That is, the cumulative amount of work done equals

$$\int_0^6 (2 + \cos(x))\,dx = (2x + \sin(x))\big|_0^6 = (2\cdot 6 + \sin(6)) - (2\cdot 0 + \sin(0)) = 12 + \sin(6)$$

$$= 11.7206 \text{ foot-pounds.}$$

Example 4.5.5 During the course of a day, from 6 AM to 6 PM, solar cells on a school absorb energy at a rate given by

$$p_1(t) = \frac{75}{1 + (t - 6)^2}.$$

Here, t denotes the number of hours since 6 AM of the day in question, and p_1 is in kilowatts. During this same time period, the school uses energy at a roughly constant rate, given by $p_2(t) = 15$. The units here are the same as for p_1.

(i) Set up an integral that measures net energy generated (meaning energy absorbed minus energy used), over the first T hours of the 12-hour period in question.
(ii) Using what you know about the arctangent function, together with some guessing and checking, evaluate the integral.

(iii) Over the entire 12-hour period, does the school experience a net gain, or a net loss, of energy?

Solution. (i) Since energy comes in at a rate of $p_1(t)$, and goes out at a rate of $p_2(t)$, net energy is generated at a rate of $p(t) = p_1(t) - p_2(t)$. (Again, $p_1(t)$, $p_2(t)$, and $p(t)$ are *rates* of energy absorption/usage, so they are measures of *power*.)

Let $E(T)$ denote the net energy generated, from $t = 0$ to $t = T$. Then $E(T)$ is "the continuous sum, from $t = 0$ to $t = T$, of infinitesimal energy elements $p(t)\, dt$." That is,

$$E(T) = \int_0^T p(t)\, dt = \int_0^T \left(\frac{75}{1 + (t - 6)^2} - 15 \right) dt. \tag{4.29}$$

(ii) The above integral looks a bit daunting, but perhaps less so if we recall that $d[\arctan(t)]/dt = 1/(1 + t^2)$. That is, $1/(1 + t^2)$ has antiderivative $\arctan(t)$. Using this fact, we might guess that $75/(1 + (t - 6)^2)$ has antiderivative $75 \arctan(t - 6)$. And we'd be right:

$$\frac{d}{dt}[75 \arctan(t - 6)] = 75 \frac{d}{dt}[\arctan(t - 6)] = 75 \cdot \frac{1}{1 + (t - 6)^2} \frac{d}{dt}[t - 6] = \frac{75}{1 + (t - 6)^2}.$$

So (using also the fact that 15 has antiderivative $15t$), Eq. (4.29) gives

$$E(T) = \int_0^T \left(\frac{75}{1 + (t - 6)^2} - 15 \right) dt$$

$$= (75 \arctan(t - 6) - 15t)\Big|_0^T = (75 \arctan(T - 6) - 15T) - (75 \arctan(0 - 6) - 15 \cdot 0)$$

$$= 75 \arctan(T - 6) - 15T - 75 \arctan(-6)$$

killowatt-hours.

(iii) We have

$$E(12) = 75 \arctan(12 - 6) - 15 \cdot 12 - 75 \arctan(-6) = 30.8471 \text{ killowatt-hours.}$$

Since net energy generated is positive, the school generates more energy than it uses, over the 12-hour period.

As Fig. 4.14 illustrates, $p_1(t)$ is sometimes larger, and sometimes smaller, than $p_2(t)$. So this example demonstrates, among other things, how accumulation functions and The Fundamental Theorem of Calculus can work even when the function being "accumulated" (in this case, $p(t) = p_1(t) - p_2(t)$) sometimes assumes negative values.

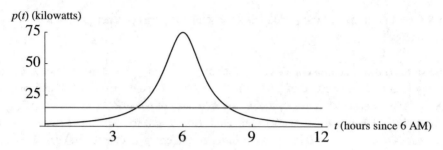

Fig. 4.14 Power generated (in black) and consumed (in red)

Example 4.5.6 Find the volume V of a square pyramid of height h and baselength b. (If you already know the formula for the volume of a square pyramid, pretend for the time being that you don't.)

Solution. We situate the pyramid with its apex at the origin, and with the positive x-axis intersecting the pyramid's base at its center.

We can imagine that the pyramid is made up of infinitesimally thin "slices," like the one shown in red in Fig. 4.15. If such a slice has cross-sectional area $A(x)$ and infinitesimal thickness dx, then we have

$$V = \int_0^b A(x)\, dx.$$

We need a formula for $A(x)$. To deduce one, consider the pyramid viewed from the side (that is, along an axis perpendicular to the xy-plane) (Fig. 4.16).

The line segment seen at the top edge of the pyramid, from this perspective, has endpoints $(0, 0)$ and $(h, b/2)$ (since the pyramid has height h and baselength b), so this segment lies on the line

$$y = \frac{b}{2h}x.$$

But the upper endpoint of the line segment shown in red also lies on this line; therefore, that endpoint has coordinates $(x, bx/(2h))$. The length of the red line segment then equals bx/h. But this line segment is one side of our square slice of cross-sectional area $A(x)$. So

$$A(x) = \left(\frac{bx}{h}\right)^2 = \frac{b^2}{h^2} \cdot x^2.$$

Finally, then,

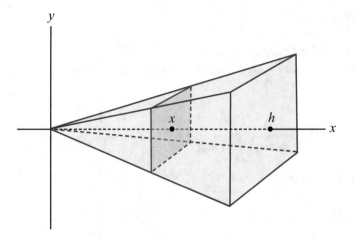

Fig. 4.15 A square pyramid positioned along the x-axis

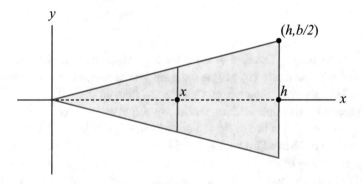

Fig. 4.16 Side view of the pyramid

$$V = \int_0^h A(x)\,dx = \int_0^h \frac{b^2}{h^2} \cdot x^2\,dx = \frac{b^2}{h^2} \int_0^h x^2\,dx$$

$$= \frac{b^2}{h^2} \cdot \frac{x^3}{3}\Big|_0^h = \frac{b^2}{h^2}\left(\frac{h^3}{3} - \frac{0^3}{3}\right) = \frac{b^2 h}{3},$$

which agrees with a known formula that you may have seen in high school geometry.

4.5.4 Exercises

4.5.4.1 Part 1: Formulas for Integrals

1. Determine the exact value of each of the following integrals.

(a) $\int_0^1 dt$

(Note that $\int_0^1 dt$ denotes $\int_0^1 1\,dt$.)

(b) $\int_3^7 (2 - 3x + 5x^2)\,dx$

(c) $\int_0^{5\pi} \sin x\,dx$

(d) $\int_0^{5\pi} \sin(2x)\,dx$

(e) $\int_0^1 e^t\,dt$

(f) $\int_1^6 \dfrac{dx}{x}$

(Note that $\dfrac{dx}{x}$ denotes $\dfrac{1}{x}\,dx$.)

(g) $\int_0^4 (7u - 12u^5)\,du$

(h) $\int_0^1 2^t\,dt$

(i) $\int_{-1}^1 s^2\,ds$

(j) $\int_{-1}^1 s^3\,ds$

2. Use the Fundamental Theorem of Calculus to compute each of the two integrals of Example 4.4.2. Make sure you get the same answers as were obtained in that example.

3. Use the Fundamental Theorem of Calculus to find $\int_1^6 \sqrt[3]{x-1}\,dx$. Hint: following the approach of Example 4.5.3(d) above, we might guess that an antiderivative of $\sqrt[3]{x-1} = (x-1)^{1/3}$ is $(x-1)^{4/3}$. Check whether this antiderivative really works, by differentiating it. Then adjust your guess if necessary.

4. Express the values of the following integrals in terms of any parameters they contain. Simplify as much as possible. You may need to do some guessing and checking to find the appropriate antiderivatives.

    Example:

    $$\int_0^1 e^{bx}\,dx = \frac{e^{bx}}{b}\bigg|_0^1 = \frac{e^{b\cdot 1}}{b} - \frac{e^{b\cdot 0}}{b} = \frac{e^b - 1}{b}.$$

(a) $\int_3^7 kx\,dx$

(b) $\int_0^\pi \sin(\alpha x)\,dx$

(c) $\int_1^4 (px^2 - x^3)\,dx$

(d) $\int_0^1 e^{ct}\,dt$

(e) $\int_{\ln 2}^{\ln 3} e^{ct}\,dt$

(f) $\int_1^b (5 - x)\,dx$

(g) $\int_0^1 a^t\,dt$

(h) $\int_1^2 u^c\,du$

4.5.4.2 Part 2: The Fundamental Theorem of Calculus and Accumulation Functions

5. A space heater runs over a 24-hour period, at a power level of

$$f(t) = 1000\left(1 + \cos\left(\frac{\pi}{12}t\right)\right)$$

   watts, where t is the number of hours since midnight (which is $t = 0$).

   (a) Write down an integral to measure how much energy this heater uses over the course of the 24-hour period (from midnight to midnight).
   (b) Evaluate the integral in part (a), to four decimal places. (Hint: An antiderivative of $\cos(kx)$ is $\sin(kx)/k$, if k is a nonzero constant.) What are the units for your answer?

6. Consider the car of Example 4.2.2. Suppose the velocity function $v(t)$ depicted there has formula

$$v(t) = 127t - 90t^2 + 17.35t^3 + 5t^4 - 2.079t^5 + 0.18t^6.$$

   Use the Fundamental Theorem of Calculus to find the distance traveled by the car, over the 5-hour period. Check your answer against the estimate we obtained in that example, to make sure you're in the right ballpark.

7. Consider a ball thrown straight up into the air. Suppose the ball has velocity (=rate of change of height with respect to time) given by $v(t) = -32t + 72$, where t denotes the number of seconds elapsed from the time the ball is released, and $v(t)$ is measured in feet per second. Then if $v(t)$ is positive, the ball is moving upwards (because $v(t) > 0$ means the height of the ball above the ground is *increasing*), while if $v(t)$ is negative, the ball is moving downwards (since then the height of the ball above the ground is *decreasing*).

   Suppose that, the moment the ball is released, it is at a height of 7 ft.

   (a) What is the height h of the ball above the ground after 3 s? After 6 s? After T seconds? Hint: the integral of $v(t)$ measures *change* in height. To get actual height $h(T)$ from this, you need to add the initial height of 7 ft.
   (b) At what time t does the ball's velocity equal zero? What does this signify in terms of the ball's motion?
   (c) What's the maximum height attained by the ball?
   (d) How long does the ball take to reach the ground? Hint: Solve $h(T) = 0$ for T. (Use the quadratic formula; it gives you two solutions. Which one can you ignore?)

8. You have a large circular kiddie pool in your backyard. It has radius 10 ft, so that the area of the base of the pool is $\pi \cdot 10^2 = 100\pi$ square feet. The pool is 8 ft deep. You are using a hose from your house to fill it to the top.
   Please supply the appropriate units with all of your answers below.

   (a) (Note: you should be able to do this part of the exercise without any calculus.) Your hose supplies 10π cubic feet of water an hour (it's one of those fancy new π-hoses from late night TV).

   (i) What will the *depth* of the water be after 1 h? after 2 h?
   (ii) What will the depth of the water be after T hours?
   (iii) How long will it take to fill the pool to 8 ft?

   (b) (Note: you probably **do** need calculus for this part of the exercise.) This is taking too long, so you buy the new Super Hose (only $19.95 plus shipping and handling, with a free Ginsu Knife if you order before midnight tonight!), which supplies water at rate given by $w(t) = 10\pi t$ cubic feet of water per hour in hour t. (The longer the hose it on, the faster the water flows out of it!)

   (i) What will the *volume* of water in the pool be after T hours? Your answer should be a function of T, call it $V(T)$.
   (ii) What will the *depth* of the water in the pool be after T hours? Your answer should be a function of T, call it $D(T)$.
   (iii) What will the depth of the water be after 1 h? after 2 h?
   (iv) How long will it take to fill the pool to 8 ft?
   (v) What is the rate of change of $D(T)$?

9. A pyramid is 30 ft tall. The area of a horizontal cross-section x feet from the top of the pyramid measures $2x^2$ square feet. What is the area of the base? What is the volume of the pyramid, to the nearest cubic foot? (See Example 4.5.6, above.)

10. What is the volume of a right circular cone of height h and radius r?

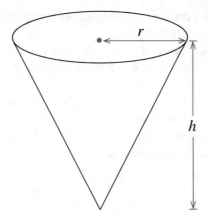

(Use an approach like that of Example 4.5.6, above.)

11. What is the volume of the "solid of revolution" you get by revolving the curve given by $y = x^2$, from $x = 0$ to $x = b$ (where $b > 0$), around the x-axis? Hint: use ideas from Example 4.5.6, above. Also note that a point (x, x^2) on this curve traces out a circle as it is revolved around the x-axis. What is the area $A(x)$ of the circle thus obtained?

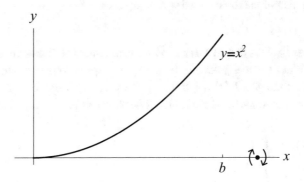

12. Repeat Exercise 11 above for the curve given by $y = \sin(x)$, from $x = 0$ to $x = \pi$. Hint: the identity

$$\sin^2(x) = \frac{1}{2}\left(1 - \cos(2x)\right)$$

may be useful at the point when you need to apply the Fundamental Theorem of Calculus.

13. (a) Repeat Exercise 11 above for the curve given by $y = 1/x$, from $x = 1$ to $x = b$ ($b > 0$).

(b) Show that your answer to part (a) of this exercise tends to a finite number as $b \to \infty$. What is this number? Based on this, do you think solids of infinite extent can have finite volume? Please explain.

4.5.4.3 Part 3: The Fundamental Theorem of Calculus for a Piecewise Continuous Function

14. Verify that the conclusion of the Fundamental Theorem of Calculus holds, with $a = 0$ and $b = 8$, for the functions E and p of Example 4.1.4 above, even though:

(i) The function p there is merely piecewise continuous, and not continuous, on the interval $[0, 8]$ (it's discontinuous at $T = 3.5$ and $T = 5$), and

(ii) there are (exactly two) points in $(0, 8)$ where $E'(T)$ *fails* to equal $p(T)$ (since $E'(T)$ fails to exist at $T = 3.5$ and $T = 5$).

Hints:

(a) Use Eq. (4.3) to evaluate $E(8) - E(0)$.

(b) We have

$$\int_0^8 p(t)\, dt = \int_0^{3.5} p(t)\, dt + \int_{3.5}^5 p(t)\, dt + \int_5^8 p(t)\, dt.$$

The three integrals on the right hand side are easily evaluated, since $p(t)$ is constant on each of the three domains of integration.

4.5.4.4 Part 4: The Other Part of The Fundamental Theorem of Calculus

15. Find $F'(x)$ for the following functions, using the following strategy. First, evaluate the given integral $F(x)$ using the Fundamental Theorem of Calculus. The result will depend on x, since the upper limit of integration does. Then, differentiate $F(x)$ directly.

(a) $F(x) = \displaystyle\int_0^x (t^2 + t^3)\, dt.$

(b) $F(x) = \displaystyle\int_1^x \frac{1}{u}\, du.$

(c) $F(x) = \displaystyle\int_0^x e^v\, dv.$

(d) $F(x) = \displaystyle\int_0^{\cos(x)} e^v\, dv.$

(e) $F(x) = \displaystyle\int_0^{x^2} \cos(t)\, dt.$

16. In the course of proving The Fundamental Theorem of Calculus, we saw that (under appropriate conditions) the integral

$$\int_a^T p(t)\, dt$$

has derivative equal to $p(T)$. (See **Fact 1** in Sect. 4.5.2.) That is,

$$\frac{d}{dT} \int_0^T p(t)\, dt = p(T).$$

In other references, this is often written a bit differently: replacing T by x and p by f, we get

$$\frac{d}{dx} \int_a^x f(t)\, dt = f(x). \tag{4.30}$$

This last formula is often called **The Fundamental Theorem of Calculus Part II**. (Actually, in many references, this result is called **The Fundamental Theorem of Calculus Part I**, and what we called the The Fundamental Theorem of Calculus earlier in this section is referred to as **The Fundamental Theorem of Calculus Part II**.) Use Eq. (4.30) to differentiate each of the integrals (a), (b), and (c) from Exercise 15 directly, without having to first perform the integration.

17. **Differentiating integrals whose upper limits of integration are themselves functions.** Use Eq. (4.30) above, together with the chain rule, to show that, under appropriate conditions on the functions f and g, we have

$$\frac{d}{dx} \int_a^{g(x)} f(t)\, dt = f(g(x))\, g'(x). \tag{4.31}$$

18. Use Eq. (4.31) to differentiate each of the integrals (d) and (e) from Exercise 15 directly, without having to first perform the integration.

19. Use Eq. (4.30) or Eq. (4.31) to find the derivative of each of the following functions F.

(a) $F(x) = \displaystyle\int_1^x \ln(1 + t^2)\, dt.$

(b) $F(x) = \displaystyle\int_0^x \cos(t^2)\, dt.$

(c) $F(x) = \displaystyle\int_0^{x^2} \cos(t^2)\, dt.$

(d) $F(x) = \displaystyle\int_1^{e^x} \frac{v}{1 + v^3}\, dv.$

(e) $F(x) = \displaystyle\int_1^{e^{\sin(x)}} \frac{v}{1 + v^3}\, dv.$

4.6   Antiderivatives

The Fundamental Theorem of Calculus gives us a way to find an integral using antiderivatives. It will be helpful, then, to explore the notion of an antiderivative a bit more closely.

Again, we say that F is an **antiderivative** of f on and interval (a, b) if $F' = f$ there. Here are some examples (Table 4.1).

All of these antiderivatives are easily verified, by differentiating the function F on the right-hand side of any given row, and checking that you get the corresponding function on the left. For example,

$$\frac{d}{dx}\left[-\frac{\cos(ax)}{a}\right] = -\frac{1}{a} \cdot \frac{d}{dx}[\cos(ax)] = -\frac{1}{a} \cdot (-a\sin(ax)) = \sin(ax),$$

which verifies the second row of the table.

The third-to-last row of the table merits some explanation. It would be incorrect to say that $\ln(x)$ is always an antiderivative of $1/x$: for one thing, $\ln(x)$ is not even defined for $x < 0$, while $1/x$ is. We'd like to find a function F such that $F'(x) = 1/x$ whenever $1/x$ makes sense. We claim that $F(x) = \ln(|x|)$ does the job. To see this, note first that $F(x) = \ln(|x|)$

Table 4.1 A short table of antiderivative formulas

Function f	Antiderivative F
$x^p \quad (p \neq -1)$	$\dfrac{x^{p+1}}{p+1}$
$\sin(ax) \quad (a \neq 0)$	$-\dfrac{\cos(ax)}{a}$
$\cos(ax) \quad (a \neq 0)$	$\dfrac{\sin(ax)}{a}$
$e^{ax} \quad (a \neq 0)$	$\dfrac{e^{ax}}{a}$
b^x	$\dfrac{b^x}{\ln(b)}$
$\dfrac{1}{x}$	$\ln(|x|)$
$\dfrac{1}{\sqrt{1-x^2}}$	$\arcsin(x)$
$\dfrac{1}{1+x^2}$	$\arctan(x)$

does make sense for all $x \neq 0$, just as $1/x$ does. Next, we differentiate $F(x)$ by considering two cases:

(i) If $x > 0$, then $|x|$ is the same as x, so

$$\frac{d}{dx}[\ln(|x|)] = \frac{d}{dx}[\ln(x)] = \frac{1}{x},$$

as desired. Next,

(ii) If $x < 0$, then $|x|$ is the same as $-x$, so

$$\frac{d}{dx}[\ln(|x|)] = \frac{d}{dx}[\ln(-x)] = \frac{1}{-x}\frac{d}{dx}[-x] = \frac{1}{-x} \cdot (-1) = \frac{1}{x}.$$

Together, (i) and (ii) tell us that $d[\ln(|x|)]/dx = 1/x$ whenever $x \neq 0$, and this confirms the third-to-last row of our table.

The last two rows of the table are implied by the differentiation formulas

$$\frac{d}{dx}[\arcsin(x)] = \frac{1}{\sqrt{1 - x^2}} \quad \text{and} \quad \frac{d}{dx}[\arctan(x)] = \frac{1}{1 + x^2},$$

which were derived in Sect. 3.3. (See Example 3.3.2, and Exercise 11, Sect. 3.3.1.)

While a function can have only one derivative, it has many antiderivatives. For example, the functions $1 - \cos(u)$ and $99 - \cos(u)$ are also antiderivatives of the function $\sin(u)$, since

$$\frac{d}{du}[1 - \cos(u)] = \sin(u) = \frac{d}{du}[99 - \cos(u)].$$

In fact, every function $F(u) = C - \cos(u)$ is an antiderivative of $f(u) = \sin(u)$, for any constant C whatsoever. This observation is true in general. That is, if F is an antiderivative of a function f, then so is $F + C$, for any constant C. This follows from the addition rule for derivatives, because if $F' = f$, then

$$(F + C)' = F' + C' = F' + 0 = f;$$

that is, $(F + C)' = f$ as well.

Remark 4.6.1 It is tempting to claim the converse—that *every* antiderivative of f is equal to $F + C$, for some appropriately chosen value of C. In fact, you will often see this statement written. The statement is true, though, only for continuous functions with no breaks in their domains. If the domain of f does have breaks, then there will be more antiderivatives than those of the form $F + C$ for a *single* constant C. Instead, over each "unbroken" piece of the domain of f, F can be modified by a *different* constant and still yield an antiderivative for f. Exercises 11 and 12 at the end of this section explore this for a couple of cases. If f is continuous on an interval, though, $F + C$ will cover all the possibilities, and we sometimes say that $F + C$ is *the* antiderivative of f. For the sake of keeping a compact notation, we will even write this when the domain of f has breaks. But you should understand that, in such cases, over each piece of this domain F can be modified by a different constant.

4.6.1   Notation

As we've noted previously—see Example 4.5.3(e)—there are functions that, even though they are expressible in terms of familiar quantities, do not have antiderivatives that can be written in closed form. This is not necessarily to say that these functions don't *have* antiderivatives.

To see this, let $p(t)$ be any sufficiently reasonable function on an interval $[a, b]$. (For example, a function that is continuous on this interval is reasonable enough.) Consider the accumulation function

$$E(T) = \int_a^T p(t) \, dt.$$

As we've seen in Sect. 4.5 (see **Fact 1** in Sect. 4.5.2), we have $E'(T) = p(T)$ for any number T between a and b. That is: *E is an antiderivative of p* for such values of T.

The point is that the definite integral gives us a means of *defining* antiderivatives. For example,

$$F(T) = \int_0^T e^{-t^2/2} \, dt \tag{4.32}$$

is an antiderivative of $f(t) = e^{-t^2/2}$, even though, again, there's no "closed" formula for this antiderivative. (The formula (4.32) is not considered "closed" because it requires the integral sign.)

The connection between antiderivatives and integrals is so pervasive that the integral sign—with the "limits of integration" omitted—is also used to denote an antiderivative:

Notation : The most general antiderivative of f is denoted $\displaystyle\int f(x) \, dx$.

Remark 4.6.2 By "the most general antiderivative of f," we mean "the *set* of all possible antiderivatives of f." So strictly speaking, $\int f(x) \, dx$ denotes not a single function, but a set of functions. Generally speaking, though, we can find all elements of this set by just finding one element, and then adding an "arbitrary constant" $+C$ to that single element. See Remark 4.6.1 above.

With this new notation, the antiderivatives we have listed so far can be written in the following form.

$$\int x^p \, dx = \frac{x^{p+1}}{p+1} + C \quad (p \neq -1)$$

$$\int \sin(ax) \, dx = -\frac{\cos(ax)}{a} + C \quad (a \neq 0)$$

$$\int \cos(ax) \, dx = \frac{\sin(ax)}{a} + C \quad (a \neq 0)$$

$$\int e^{ax} \, dx = \frac{e^{ax}}{a} + C \quad (a \neq 0)$$

$$\int b^x \, dx = \frac{b^x}{\ln b} + C \quad (b > 0)$$

$$\int \frac{1}{x} \, dx = \ln(|x|) + C$$

$$\int \frac{1}{\sqrt{1-x^2}} \, dx = \arcsin(x) + C$$

$$\int \frac{1}{1+x^2} \, dx = \arctan(x) + C$$

The integral sign $\int$ now has two distinct meanings. Originally, it was used to describe the *number*

$$\int_a^b f(x) \, dx,$$

which is a signed area, or a limit of a sequence of Riemann sums. Because this integral has a definite numerical value, it is called the **definite integral**. In its new meaning, the integration sign is used to describe the antiderivative

$$\int f(x) \, dx,$$

which is a *function* (really, a *set* of functions), not a number. To contrast the new use of $\int$ with the old, and to remind us that the new expression $\int f(x) \, dx$ is a variable quantity—that is, $\int f(x) \, dx$ is a function of x (or, more precisely, a function of x plus an arbitrary constant), it is called the **indefinite integral** of f.

The function that appears in either a definite or an indefinite integral is called the **integrand**.

Because an indefinite integral entails antidifferentiation, the process of finding an antiderivative is sometimes called **integration**. We've also used this term to designate the process of finding a definite integral. Thus the term *integration*, as well as the symbol for it, has two distinct meanings.

4.6.2 Using Antiderivatives

According to the Fundamental Theorem, we can use an *indefinite* integral to find the value of a *definite* integral—and this largely explains the importance of antiderivatives. In the language of indefinite integrals, the statement of the Fundamental Theorem in the box in Sect. 4.5.1 takes the following form.

$$\int_a^b f(x)\,dx = F(b) - F(a),\ \text{where}\ F(x) = \int f(x)\,dx.$$

Example 4.6.1 Find $\int_1^4 x^2\,dx$.

Solution. We have

$$\int x^2\,dx = \frac{1}{3}x^3 + C.$$

It follows that

$$\int_1^4 x^2\,dx = \left(\frac{1}{3}x^3 + C\right)\Big|_1^4 = \left(\frac{1}{3}\cdot 4^3 + C\right) - \left(\frac{1}{3}\cdot 1^3 + C\right) = \frac{64}{3} + C - \frac{1}{3} - C = 21.$$

Note that, in the above example, the two appearances of "+C" cancel each other. This cancellation will occur no matter what function we are integrating, since

$$(F(x) + C)\big|_a^b = (F(b) + C) - (F(a) + C) = F(b) + C - F(a) - C = F(b) - F(a) = F(x)\big|_a^b.$$

This implies that it does not matter which value of C we choose to do the calculation. Usually, we just take $C = 0$ (which amounts to the procedure we followed in Sect. 4.5).

Example 4.6.2 Find $\int_0^{\pi/2} \cos(t)\,dt$.

Solution. This time, the indefinite integral we need is

$$\int \cos(t)\,dt = \sin(t) + C.$$

The value of the definite integral is therefore

$$\int_0^{\pi/2} \cos(t)\,dt = \sin(t)\big|_0^{\pi/2} = \sin(\pi/2) - \sin(0) = 1 - 0 = 1.$$

4.6.3 Finding Antiderivatives

What we have seen above is this:

> **The statement $F'(x) = f(x)$ is the same**
>
> **as the statement** $\displaystyle\int f(x)\,dx = F(x) + C.$

Relationship between indefinite integrals and derivatives

Because of this fact, we can verify many statements about antidifferentiation by considering the corresponding differentiation facts.

In particular, the following basic indefinite integral rules can be verified using analogous rules for derivatives.

$$\int k f(x)\,dx = k \int f(x)\,dx \qquad \textbf{(constant multiple rule)};$$

$$\int (f(x) + g(x))\,dx = \int f(x)\,dx + \int g(x)\,dx \qquad \textbf{(sum rule)}.$$

(Here, k is a constant.)

For example, the sum rule for indefinite integrals may be demonstrated as follows. Let F be an antiderivative for f, and G an antiderivative for g. Then by the above boxed statement,

$$\int f(x)\,dx + \int g(x)\,dx = (F(x) + C_1) + (G(x) + C_2) = F(x) + G(x) + C, \quad (4.33)$$

where C_1 and C_2 are arbitrary constants, and $C = C_1 + C_2$. (Since C_1 and C_2 can be anything, so can C; so C is an arbitrary constant as well.) But by the sum rule for differentiation, the derivative of the right-hand side of (4.33) equals $F'(x) + G'(x) = f(x) + g(x)$, so again by the above boxed statement,

$$\int (f(x) + g(x))\,dx = F(x) + G(x) + C. \qquad (4.34)$$

The right-hand sides of Eqs. (4.33) and (4.34) are equal, so the left-hand sides are equal too, and this is what we wanted to show.

Example 4.6.3 This example illustrates the use of both the sum and the constant multiple rules.

$$\int (7e^x + \cos(x)) \, dx = \int 7e^x \, dx + \int \cos(x) \, dx$$

$$= 7 \int e^x \, dx + \int \cos(x) \, dx = 7e^x + \sin(x) + C.$$

The next example illustrates how initial value problems of the form

$$F'(x) = f(x), \qquad F(0) = y_0 \tag{4.35}$$

may be solved using indefinite integrals.

Example 4.6.4 Find a function F such that

$$F'(x) = 3x^2 - \sin(x), \qquad F(0) = 7.$$

Solution. We are looking for a function F of x whose derivative is a given function of x, and whose value at a certain point is a given number. The general strategy for such problems is to first find *any and all* functions with the indicated derivative, and then to select, among all of those functions, the one that satisfies the given initial condition.

In other words, the first step is to find the *most general antiderivative*, which is to say the *indefinite integral*, of the given function. That is: the equation $F'(x) = 3x^2 - \sin(x)$ tells us that

$$F(x) = \int (3x^2 - \sin(x)) \, dx = x^3 + \cos(x) + C. \tag{4.36}$$

Now, we need only figure out what C is.

To do so, we substitute the condition $F(0) = 7$ into Eq. (4.36), to get

$$7 = F(0) = 0^3 + \cos(0) + C = 1 + C.$$

Solving $7 = 1 + C$ for C gives $C = 7 - 1 = 6$. We plug this value of C back into (4.36) to get our complete solution:

$$F(x) = x^3 + \cos(x) + 6.$$

4.6.4 Exercises

4.6.4.1 Part 1: Basic Antidifferentiation

1. Find a formula for each of the following indefinite integrals. For each integral, verify that your result is correct by differentiation.

Example:

$$\int 4\cos(3x+2)\,dx = \frac{4}{3}\sin(3x+2) + C.$$

Verification:

$$\frac{d}{dx}\left[\frac{4}{3}\sin(3x+2) + C\right] = \frac{4}{3}\frac{d}{dx}[\sin(3x+2)] + 0$$

$$= \frac{4}{3}\cdot\cos(3x+2)\cdot\frac{d}{dx}[3x+2] = \frac{4}{3}\cdot\cos(3x+2)\cdot 3 = 4\cos(3x+2).$$

(a) $\displaystyle\int 3x\,dx$

(b) $\displaystyle\int 3u\,du$

(c) $\displaystyle\int e^z\,dz$

(d) $\displaystyle\int (5t^4 + 5\cdot 4^t)\,dt$

(e) $\displaystyle\int \left(7y + \frac{1}{y}\right) dy$

(f) $\displaystyle\int \left(7y - \frac{4}{y^2}\right) dy$

(g) $\displaystyle\int (5\cos(w) - \cos(5w))\,dw$

(h) $\displaystyle\int dx$ (This just means $\displaystyle\int 1\,dx$.)

(i) $\displaystyle\int e^{z+2}\,dz$

(j) $\displaystyle\int \cos(4x - 2)\,dx$

(k) $\displaystyle\int \frac{5}{1+r^2}\,dr$

(l) $\displaystyle\int \frac{1}{1+4s^2}\,ds$ (Hint: guess and check,

using the fact that $\dfrac{d}{ds}[\arctan(s)] = \dfrac{1}{1+s^2}$.)

(m) $\displaystyle\int (2x + 3)^7\,dx$

(n) $\displaystyle\int \cos(1 - x)\,dx$

2. Find $\displaystyle\int (a + by)\,dy$, where a and b are constants.

4.6.4.2 Part 2: Initial Value Problems

3. (a) Solve the initial value problem

$$F'(x) = 7, \qquad F(0) = 12.$$

(b) Solve the initial value problem

$$G'(x) = 7, \qquad G(3) = 1.$$

(c) Do $F(x)$ and $G(x)$ differ by a constant? If so, what is the value of that constant?

4. (a) Find an antiderivative $F(t)$ of $f(t) = t + \cos(t)$ for which $F(0) = 3$.

(b) Find an antiderivative $G(t)$ of $f(t) = t + \cos(t)$ for which $G(\pi/2) = -5$.

(c) Do $F(t)$ and $G(t)$ differ by a constant? If so, what is the value of that constant?

5. Solve the initial value problem

$$\frac{dy}{dx} = \frac{1}{1+x^2}, \qquad y(0) = 4.$$

6. Solve the initial value problem

$$\frac{dp}{dq} = 4q - \frac{3}{q^2}, \qquad p(1) = 6.$$

4.6.4.3 Part 3: Guessing and Checking

For part (a) of each of the exercises below, note that the instruction "Verify that $F(x)$ is an antiderivative of $f(x)$" simply means "show that $F'(x) = f(x)$."

7. (a) Verify that $(1+x^3)^{10}$ is an antiderivative of $30x^2(1+x^3)^9$.
   (b) Find an antiderivative of $x^2(1+x^3)^9$.
   (c) Find an antiderivative of $x^2 + x^2(1+x^3)^9$.
8. (a) Verify that $x \ln(x)$ is an antiderivative of $1 + \ln(x)$.
   (b) Find an antiderivative of $\ln(x)$. (Do you see how you can use part (a) to find this antiderivative?)
9. Recall that $F(y) = \ln(y)$ is an antiderivative of $1/y$ for $y > 0$. According to the text, *every* antiderivative of $1/y$ over this domain must be of the form $\ln(y) + C$ for an appropriate value of C.

   (a) Verify that $G(y) = \ln(2y)$ is also an antiderivative of $1/y$.
   (b) Find C so that $\ln(2y) = \ln(y) + C$.

10. (a) Verify that $-\cos^2(t)$ is an antiderivative of $2\cos(t)\sin(t)$.
    (b) Since $\sin^2(t)$ is an antiderivative of $2\cos(t)\sin(t)$ (as you can verify directly), you should be able to show that

$$-\cos^2(t) = \sin^2(t) + C$$

   for an appropriate value of C. What is C?

4.6.4.4 Part 4: Miscellaneous

11. The function $\ln(|x|) + C$ is an antiderivative of $1/x$, for any constant C, but there are more antiderivatives. This can happen because the domain of $1/x$ is broken into two parts. To see this, let

$$G(x) = \begin{cases} \ln(-x) & \text{if } x < 0, \\ \ln(x) + 1 & \text{if } x > 0. \end{cases}$$

(a) Explain why there is no value of C for which

$$\ln(|x|) + C = G(x).$$

This shows that the functions $\ln(|x|) + C$ do not exhaust the set of antiderivatives of $1/x$.

(b) Construct two more antiderivatives of $1/x$ and sketch their graphs. What is the general form of the new antiderivatives you have constructed? (A suggestion: you should be able to use two separate constants C_1 and C_2 to describe the general form.)

12. In the list in Sect. 4.6.1, the antiderivative of x^p is given as

$$\frac{1}{p+1} x^{p+1} + C.$$

For some values of p this is correct, with only a single constant C needed. For other values of p, though, the domain of x^p will consist of more than one piece, and $\frac{1}{p+1} x^{p+1}$ can be modified by a different constant over each piece. For what values of p does this happen?

13. Consider the two functions

$$F(x) = \sqrt{1 + x^2} - 1 \quad \text{and} \quad G(x) = \int_0^x \frac{t}{\sqrt{1+t^2}}\, dt.$$

(a) Show that F and G both satisfy the initial value problem

$$y' = \frac{x}{\sqrt{1+x^2}}, \qquad y(0) = 0.$$

(b) Since an initial value problem typically has a *unique* solution, F and G should be equal. Assuming this, determine the exact value of the following definite integrals.

$$\int_0^1 \frac{t}{\sqrt{1+t^2}}\, dt, \quad \int_0^2 \frac{t}{\sqrt{1+t^2}}\, dt, \quad \int_0^5 \frac{t}{\sqrt{1+t^2}}\, dt.$$

14. Find the area under the curve $y = x^3 + x$ for x between 1 and 4.
15. Find the area under the curve $y = e^{3x}$ for x between 0 and $\ln 3$.

4.7   Integration by Substitution

In the preceding section, we reimagined a couple of general rules for differentiation—the constant multiple rule and the sum rule—in integral form. In this section we will develop the integral form of the chain rule, and see some of the ways this can be used to evaluate indefinite and definite integrals.

4.7.1   Substitution in Indefinite Integrals

We begin with the following.

Example 4.7.1 Evaluate the indefinite integral

$$\int 2x \cos(x^2)\,dx. \tag{4.37}$$

Solution. The key fact here is that the integrand contains a quantity—namely, x^2—whose *derivative*—namely, $2x$—also appears as a factor in the integrand. Let's see how we can make use of this fact.

The strategy in cases like this is to give a name to the quantity whose derivative is present. We call that quantity u; that is,

$$u = x^2. \tag{4.38}$$

Then, of course, we have $du/dx = x^2$ or, "multiplying both sides by dx,"

$$du = 2x\,dx. \tag{4.39}$$

Of course we can't *really* "multiply both sides by dx," because a derivative is not actually a fraction—du and dx are not separate quantities, but rather, they are just pieces of the symbol du/dx. But let's proceed as if this strategy were justifiable—it *is* in fact justifiable, and we'll explain why at the end of this subsection.

The key thing to observe, to complete our problem, is this: if we substitute (4.38) and (4.39) into our integral (4.37), then that integral becomes a *simpler* one, involving the variable u (and only the variable u). And this simpler integral is straightforward to evaluate. Specifically, we have

$$\int 2x \cos(x^2)\,dx = \int \underbrace{\cos(x^2)}_{\substack{\text{this becomes}\\\cos(u)}} \cdot \underbrace{2x\,dx}_{\substack{\text{this is}\\\text{just } du}} = \int \cos(u)\,du = \sin(u) + C. \tag{4.40}$$

The first step here was not really necessary; we just reordered the factors in the integrand to make it more clear that, upon making the substitutions (4.38) and (4.39), this integrand (including the dx) becomes $\cos(u)\,du$.

We next observe that, in (4.40), the "question" was in terms of the variable x, while the answer was in terms of u. To be consistent, we should answer in terms of the question's original variable. This is easy: recall that $u = x^2$, so (4.40) reads

$$\int 2x\cos(x^2)\,dx = \sin(x^2) + C. \tag{4.41}$$

Strictly speaking, we're done, though it's a good idea to check our work, which we do as follows:

$$\frac{d}{dx}\left[\sin(x^2) + C\right] = \cos(x^2)\cdot\frac{d}{dx}\left[x^2\right] + 0 = 2x\cos(x^2), \tag{4.42}$$

so our solution (4.41) is correct. Note how the chain rule arises in our check. It's *because* of the chain rule that integration by substitution works. (See the discussion at the end of this subsection.)

The strategy that we employed above, and which works in many similar situations, is as follows:

Step 1. If possible, identify a quantity $g(x)$ in the integrand such that the derivative $g'(x)$ also appears as a factor in that integrand.
Step 2. Call the original quantity u. So $u = g(x)$, $du/dx = g'(x)$, and $du = g'(x)\,dx$.
Step 3. In the original integral, replace $g(x)$ by u and $g'(x)\,dx$ by du, to arrive, if possible, at an integral involving only the variable u.
Step 4. If possible, evaluate the integral in u.
Step 5. Into the answer from Step 4, replace u by $g(x)$, to put the solution in terms of the original variable.

Strategy for substitution in indefinite integrals

Here are some further examples, some of which entail slight variations on the above strategy.

Example 4.7.2 (a) Evaluate the following indefinite integrals.

(i) $\displaystyle\int (2+5x^4)(2x+x^5)^{26}\,dx$ (ii) $\displaystyle\int e^{\sin(x)}\cos(x)\,dx$ (iii) $\displaystyle\int \frac{e^x}{1+(e^x+4)^2}\,dx$

(iv) $\displaystyle\int \frac{\sin(\ln(z))}{z}\,dz$ (v) $\displaystyle\int x\sin(x^2+1)\,dx$ (vi) $\displaystyle\int f(g(x))g'(x)\,dx$

(b) Solve the initial value problem

$$\frac{dy}{dt} = \frac{(2\ln(t) + 3)^2}{t}, \qquad y(1) = 0.$$

Solution. (a) (i) We look for a part of the integrand whose derivative also appears as a factor there. We see that $u = 2x + x^5$ works: it gives us $du/dx = 2 + 5x^4$. This in turn gives $du = (2 + 5x^4)\,dx$, so

$$\int (2 + 5x^4)(2x + x^5)^{26}\,dx = \int u^{26}\,du = \frac{u^{27}}{27} + C = \frac{(2x + x^5)^{27}}{27} + C.$$

As a check on our work, we note that

$$\frac{d}{dx}\left[\frac{(2x + x^5)^{27}}{27} + C\right] = \frac{27(2x + x^5)^{26}}{27}\frac{d}{dx}[2x + x^5] + 0 = (2 + 5x^4)(2x + x^5)^{26},$$

as required.

For subsequent parts of this example, we'll omit the "check," though it's a good idea to include it unless and until you are completely comfortable with the substitution technique.

(ii) Here we keep track of substitutions "in the margin," a process we will follow hereafter.

$$\int e^{\sin(x)}\cos(x)\,dx = \int e^u\,du \qquad\qquad\qquad u = \sin(x)$$

$$= e^u + C \qquad\qquad\qquad\qquad\qquad \frac{du}{dx} = \cos(x)$$

$$= e^{\sin(x)} + C. \qquad\qquad\qquad\qquad\quad du = \cos(x)\,dx$$

(iii) As we illustrate here, one can go straight from "$u = \ldots$" to "$du = \ldots$," skipping the "$du/dx = \ldots$" step in between.

$$\int \frac{e^x}{1 + (e^x + 4)^2}\,dx = \int \frac{1}{1 + u^2}\,du \qquad\qquad u = e^x + 4$$

$$= \arctan(u) + C \qquad\qquad\qquad\qquad du = e^x\,dx$$

$$= \arctan(e^x + 4) + C.$$

One other thing to note from the above example is the following. We could have put $u = e^x$ instead of $u = e^x + 4$; we would have wound up with the same du. Generally speaking, though, we want to "break off" as much as possible into the quantity we call u, to obtain a u-integral that is as simple as possible.

(iv)

$$\int \frac{\sin(\ln(z))}{z}\, dz = \int \sin(u)\, du \qquad\qquad u = \ln(z)$$

$$= -\cos(u) + C \qquad\qquad du = \frac{1}{z}\, dz$$

$$= -\cos(\ln(z)) + C.$$

(v) If your u-substitution gives you a du that's "off by a constant factor" from where you want it to be, you can always divide through by this factor, and proceed as usual. The following example illustrates how this works.

$$\int x \sin(x^2 + 1)\, dx = \int \sin(u)\left(\frac{du}{2}\right) \qquad\qquad u = x^2 + 1$$

$$= \frac{1}{2}\int \sin(u)\, du = -\frac{1}{2}\cos(u) + C \qquad\qquad du = 2x\, dx$$

$$= -\frac{1}{2}\cos(x^2 + 1) + C. \qquad\qquad \frac{du}{2} = x\, dx$$

The point of the last line in the margin is that our substitution gives us a $2x\, dx$, while our original integral only involves an $x\, dx$. So we divide the equation $du = 2x\, dx$ by two, to get something (namely, $x\, dx$) that we can substitute directly into the integral. We then use the constant multiple rule for indefinite integrals to pull the factor of $1/2$ out front.

(vi)

$$\int f(g(x))g'(x)\, dx \qquad\qquad u = g(x)$$

$$= \int f(u)\, du. \qquad\qquad du = g'(x)\, dx$$

The above example encapsulates the general idea behind integration by substitution. Of course, one must still perform the integration in u (if possible).

(b) We first antidifferentiate:

$$y = \int \frac{(2\ln(t) + 3)^2}{t}\, dt = \int u^2\left(\frac{du}{2}\right) \qquad\qquad u = 2\ln(t) + 3$$

$$= \frac{1}{2}\int u^2\, du = \frac{1}{2}\cdot\frac{1}{3}u^3 + C \qquad\qquad du = \frac{2}{t}\, dt$$

$$= \frac{1}{6}(2\ln(t) + 3)^3 + C. \qquad\qquad \frac{du}{2} = \frac{1}{t}\, dt$$

Into this solution, we now substitute $y(1) = 0$; we get

$$0 = y(1) = \frac{1}{6}(2\ln(1) + 3)^3 + C = \frac{3^3}{6} + C = \frac{9}{2} + C.$$

Solving for C gives $C = -9/2$, so our solution is

$$y = \frac{1}{6}(2\ln(t) + 3)^3 - \frac{9}{2}.$$

It should be noted that substitution doesn't always work. For example, consider the indefinite integral

$$\int e^{-t^2/2}\, dt.$$

We mentioned earlier (see Example 4.5.3(e)) that $f(t) = e^{-t^2/2}$ does not have a closed-form antiderivative. Not knowing this, or not believing it, we might try to integrate by substitution, putting $u = -t^2/2$, for example. The problem with this approach is that it gives us $du = -t\, dt$. And what do we do with the factor of t here? We could try dividing through by $-t$; that is, we could try writing $du/(-t) = dt$. We'd get

$$\int e^{-t^2/2}\, dt = \int e^u \left(\frac{du}{-t}\right),$$

but then we're stuck with the t in the denominator. (We can't pull the t outside of the integral; the constant multiple rule applies only to *constant* multiples, but $u = -t^2$, so t is not a constant with respect to u.)

Of course, the substitution approach *does* work in many situations—situations where some function *and* its derivative appear in the integrand. And we are still left with the question: when this approach works, *why* does it work? We've not yet answered this question because, in our arguments and computations above, we have always gone from a statement of the form $u = g(x)$ to one of the form $du = g'(x)\, dx$. And we have only justified this informally: we have argued that $u = g(x)$ implies $du/dx = g'(x)$, which in turn implies $du = g'(x)\, dx$. But the second part of this argument is not rigorous, because an integral is not a fraction—and we have not even defined du or $g'(x)\, dx$.

So why *does* substitution work? Philosophically, it's because a derivative is a lot *like* a fraction. Indeed, since $du/dx = \lim_{\Delta x \to 0} \Delta u/\Delta x$, a derivative is a *limit* of fractions. Because of this, it behaves much like a fraction in many ways.

Mathematically, to say that substitution "works" is to say that the "new" integral that results from a substitution really is the *same* as the original integral into which this substitution was made. In other words, we are claiming that, if $u = g(x)$, then

$$\int f(g(x))g'(x)\, dx = \int f(u)\, du$$

(cf. Example 4.7.2(a)(vi) above). To show that this is in fact the case, let F be an antiderivative of f. Then we know that

$$\int f(u)\,du = F(u) + C = F(g(x)) + C. \tag{4.43}$$

If we can show that

$$\int f(g(x))g'(x)\,dx = F(g(x)) + C \tag{4.44}$$

as well, then the left-hand sides of (4.43) and (4.44) will be equal, and we'll be done.

To demonstrate (4.44), we need only show that the derivative of the right-hand side equals the integrand on the left. We do so using the chain rule:

$$\frac{d}{dx}[F(g(x)) + C] = F'(g(x))g'(x) = f(g(x))g'(x),$$

since, again, $F' = f$. This concludes our proof.

4.7.2   Substitution in Definite Integrals

Recall that a definite integral

$$\int_a^b f(x)\,dx$$

is a signed area between the graph of $y = f(x)$ and the interval $[a, b]$ on the x-axis. It's important to note that the limits of integration a and b are x-values. So, if we make a substitution $u = \ldots$ into our integral, this substitution affects these limits as well. (Think of the substitution $u = g(x)$ as transforming the interval $[a, b]$ into the interval $[g(a), g(b)]$.) We need to account for this in our computations.

We do so by noting, in our margin work, the effects of our substitution on the original limits $x = a$ and $x = b$. The following examples illustrate the main ideas.

Example 4.7.3 (i) We evaluate the definite integral

$$\int_0^{\pi/2} \frac{\cos(x)\,dx}{1 + \sin(x)}$$

by making the substitution $u = 1 + \sin(x)$, into both the integrand and the limits of integration:

$$\int_0^{\pi/2} \frac{\cos(x)\,dx}{1+\sin(x)} = \int_1^2 \frac{1}{u}\,du \qquad\qquad \left| \begin{array}{l} u = 1 + \sin(x) \end{array} \right.$$

$$= \ln(u)\Big|_1^2 = \ln(2) - \ln(1) \qquad\qquad \left| \begin{array}{l} du = \cos(x)\,dx \end{array} \right.$$

$$= \ln(2). \qquad\qquad\qquad\qquad \left| \begin{array}{l} \text{When } x = 0,\, u = 1 + \sin(0) = 1 \\ \text{When } x = \pi/2,\, u = 1 + \sin(\pi/2) = 2 \end{array} \right.$$

(ii) In discussing the reversal rule for definite integrals (see Sect. 4.4), we mentioned that, sometimes, it's useful to be able to integrate from a larger number to a smaller one (that is, to integrate "from right to left"). The following example demonstrates how such "backwards" integrals can arise through the substitution method.

$$\int_{-2}^1 x(5+x^2)^3\,dx = \int_9^6 u^3\left(\frac{du}{2}\right) \qquad\qquad \left| \begin{array}{l} u = 5 + x^2 \end{array} \right.$$

$$= \frac{1}{2}\int_9^6 u^3\,du = \frac{1}{2}\cdot\frac{u^4}{4}\Big|_9^6 \qquad\qquad \left| \begin{array}{l} du = 2x\,dx \end{array} \right.$$

$$= \frac{6^4 - 9^4}{8} = -\frac{5265}{8}. \qquad\qquad \left| \begin{array}{l} \dfrac{du}{2} = x\,dx \\ \text{When } x = -2,\, u = 5 + (-2)^2 = 9 \\ \text{When } x = 1,\, u = 5 + 1^2 = 6 \end{array} \right.$$

As mentioned earlier, The Fundamental Theorem of Calculus applies regardless of which limit of integration is the larger of the two. So we were able apply this theorem to the above integral from 9 to 6, without worrying especially about whether this integral denotes an area, or a signed area, or a "signed area in reverse."

4.7.3 Exercises

4.7.3.1 Part 1: Indefinite Integrals

1. Evaluate the following indefinite integrals using substitution. Make sure to indicate your substitution clearly. (That is: what are you calling u? What is du?)

(a) $\displaystyle\int 2y(y^2+1)^{50}\,dy$

(j) $\displaystyle\int \sin(w)\sqrt{\cos(w)}\,dw$

(b) $\displaystyle\int \sin(5z)\,dz$

(k) $\displaystyle\int \frac{\sin(\sqrt{s})}{\sqrt{s}}\,ds$

(c) $\displaystyle\int \frac{e^{\sqrt{x}}}{\sqrt{x}}\,dx$

(l) $\displaystyle\int \sqrt{3-x}\,dx$

(d) $\displaystyle\int (5t+7)^{50}\,dt$

(m) $\displaystyle\int \frac{dr}{r\ln r}$

(e) $\displaystyle\int 3u^2\sqrt[3]{u^3+8}\,du$

(n) $\displaystyle\int e^x\sin(1+3e^x)\,dx$

(f) $\displaystyle\int v^2(3+v^3)^4\,dv$

(o) $\displaystyle\int \frac{y}{1+y^2}\,dy$

(g) $\displaystyle\int \tan(x)\,dx$

(p) $\displaystyle\int \frac{w}{\sqrt{1-w^2}}\,dw$

(h) $\displaystyle\int \tan^2(x)\sec^2(x)\,dx$

(q) $\displaystyle\int \frac{1}{1+4y^2}\,dy$

(i) $\displaystyle\int \sec(x/2)\tan(x/2)\,dx$

(r) $\displaystyle\int \frac{e^w}{\sqrt{1+e^w}}\,dw$

4.7.3.2 Part 2: Definite Integrals

2. Use integration by substitution, together with The Fundamental Theorem of Calculus, to evaluate each of the following definite integrals. Express your answer to four decimal places. You can check your results using RIEMANN if you would like. But please also show all the work required for the substitution method. (What is u? What is du? What do your limits of integration become, under your substitution?).

(a) $\displaystyle\int_0^1 \frac{3s^2}{s^3+1}\,ds$

(e) $\displaystyle\int_0^1 \frac{t}{\sqrt{1+t^2}}\,dt$

(b) $\displaystyle\int_2^4 \frac{1}{x(\ln(x))^2}\,dx$

(f) $\displaystyle\int_0^1 \frac{\sin(\pi\sqrt{t})}{\sqrt{t}}\,dt$

(c) $\displaystyle\int_1^3 \frac{1}{2x+1}\,dx$ (Hint: $u=2x+1$)

(g) $\displaystyle\int_0^2 \frac{1}{1+(x^2/4)}\,dx$

(d) $\displaystyle\int_0^1 \frac{\arctan(x)}{1+x^2}\,dx$

(h) $\displaystyle\int_0^{\pi/2} \sin^3(x)\cos(x)\,dx$

4.7.3.3 Part 3: Initial Value Problems

3. (a) Find all functions $y=F(x)$ that satisfy the differential equation

$$\frac{dy}{dx}=x^2\left(1+x^3\right)^{13}.$$

(b) From among the functions $F(x)$ you found in part (a), select the one that satisfies $F(0) = 4$.

(c) From among the functions $F(x)$ you found in part (a), select the one that satisfies $F(-1) = 4$.

4. Find a function $y = G(t)$ that solves the initial value problem

$$\frac{dy}{dt} = te^{-t^2} \qquad y(0) = 3.$$

4.7.3.4 Part 4: Miscellaneous

5. This question concerns the indefinite integral $I = \displaystyle\int \sin(x)\cos(x)\,dx$.

(a) Find I by using the substitution $u = \sin(x)$.

(b) Find I by using the substitution $u = \cos(x)$.

(c) Compare your answers to (a) and (b). Are they the same? If not, how do they differ? Since both answers are antiderivatives of $\sin(x)\cos(x)$, they should differ only by a constant. Is that true here? If so, what is the constant?

(d) Now calculate the value of the *definite* integral

$$\int_0^{\pi/2} \sin(x)\cos(x)\,dx$$

twice, using the two *indefinite* integrals you found in (a) and (b). Do the two values agree, or disagree? Is your result consistent with what you expect?

6. (a) Sketch the graph of the function $y = xe^{-x^2}$ on the interval $[0, 5]$.

(b) Find the area between the graph of $y = xe^{-x^2}$ and the x-axis for $0 \le x \le 5$.

(c) Find the area between the graph of $y = xe^{-x^2}$ and the x-axis for $0 \le x \le b$. Express your answer in terms of the quantity b, and denote it $A(b)$. (So $A(5)$ is the number you found in part (b).) What are the values of $A(10)$, $A(100)$, $A(1000)$?

(d) It is possible to argue that the area between the graph of $y = xe^{-x^2}$ and the *entire* positive x-axis is $1/2$. Can you develop such an argument?

4.8   Separation of Variables

One of the principal uses of integration techniques is to find closed form solutions to differential equations. If you look back at the methods we have developed so far in this chapter, they are all applicable to differential equations of the form $y' = f(t)$ for some function f—that is, to differential equations where *the rate at which the **dependent** variable changes is a function of the **independent** variable only*. Such differential equations are often called "pure-time" equations, because they express rates of change purely in terms of the independent variable, which is frequently a time variable. See, for instance, Examples 4.6.4 and 4.7.2(b) above.

For pure-time initial value problems, we only need to find an antiderivative F for f, choose the constant C to satisfy the initial value, and we have our solution.

As we saw in earlier chapters, though, the behavior of y' often depends on the values of the *dependent* variable y, rather than the independent variable t—think of the SIR model, the exponential and logistic growth models, predator-prey problems, and so on. When y' is expressed purely in terms of y, we often call our differential equation "autonomous." In this section, we will see how our earlier techniques can be adapted to apply to autonomous equations, and to various differential equations of "mixed" type, where dy/dx is expressed in terms of both y and x.

4.8.1   The Separation of Variables Procedure

The following example illustrates the general ideas.

Example 4.8.1 Solve the initial value problem

$$\frac{dy}{dx} = xy^2, \qquad y(1) = 3. \tag{4.45}$$

Solution. In the previous section, we employed the useful "trick" of treating a derivative du/dx like a fraction, so that we could manipulate the so-called "differentials" du and dx separately. Let's see how this idea can also help us in the present context.

We begin by rearranging the differential equation in (4.45) so that all y's (including the dy) are on the left-hand side, and all x's (including the dx) are on the right. To do this, we multiply both sides of (4.45) by dx, and divide both sides by y^2, to get

$$y^{-2}\, dy = x\, dx. \tag{4.46}$$

Placing an indefinite integral sign in front of the quantity on each side of (4.46) then gives

$$\int y^{-2}\, dy = \int x\, dx. \tag{4.47}$$

We perform the indicated antidifferentiations, to get

$$-y^{-1} + C_1 = \frac{x^2}{2} + C_2, \tag{4.48}$$

where C_1 and C_2 are (perhaps different) constants. We subtract C_1 from both sides of (4.48) to get

$$-y^{-1} = \frac{x^2}{2} + C, \tag{4.49}$$

where we have denoted the constant $C_2 - C_1$ simply by C.

The next step is to solve for C. We do so by substituting the initial condition $y(1) = 3$ (which says: when $x = 1$, $y = 3$) into (4.49), to get

$$-3^{-1} = \frac{1^2}{2} + C$$

$$-\frac{1}{3} = \frac{1}{2} + C$$

$$C = -\frac{1}{3} - \frac{1}{2} = -\frac{2+3}{6} = -\frac{5}{6}.$$

Putting this back into (4.49) gives

$$-y^{-1} = \frac{x^2}{2} - \frac{5}{6} = \frac{3x^2 - 5}{6}. \tag{4.50}$$

We obtained a common denominator on the right, because this will facilitate the final step, which is to solve for y. We do so by multiplying both sides of (4.50) by -1, and then taking reciprocals:

$$y^{-1} = \frac{-3x^2 + 5}{6}$$

$$y = \frac{6}{-3x^2 + 5}. \tag{4.51}$$

In many previous examples involving antidifferentiation, we have checked our work by differentiating. We can do that here, too. However, the process is a bit more complicated now, because our derivative was given to us originally in terms of both x and y. One way to proceed, in such a case, is to express *both* sides of the given differential equation in terms of x alone, and to check that these two expressions in x really are equal.

Specifically: let y be given by (4.51). Then on the one hand, we have

$$\frac{dy}{dx} = \frac{d}{dx}\left[\frac{6}{-3x^2+5}\right]$$

$$= 6\frac{d}{dx}\left[(-3x^2+5)^{-1}\right] = 6\cdot(-(-3x^2+5)^{-2})\cdot\frac{d}{dx}\left[-3x^2+5\right] = 36x(-3x^2+5)^{-2},$$

$$(4.52)$$

while on the other hand, we have

$$xy^2 = x\left(\frac{6}{-3x^2+5}\right)^2 = \frac{36x}{(-3x^2+5)^2}. \tag{4.53}$$

The quantities in (4.52) and (4.53) are equal, so our function y of (4.51) indeed does satisfy the differential equation in (4.45).

To check that this function y also satisfies the initial condition in (4.45) is easier: we have

$$y(1) = \frac{6}{-3\cdot 1^2 + 5} = \frac{6}{2} = 3,$$

so our initial condition is satisfied, and we are done with our check.

In the above example, we followed a "separate, integrate, evaluate, solve" strategy. This is a general approach, which we will summarize below, to solving initial value problems of the form

$$\frac{dy}{dx} = f(x)g(y), \qquad y(x_0) = y_0. \tag{4.54}$$

Note that the derivative in (4.54) takes a very special form: it's equal to a function of x *times* a function of y. If the derivative is not of that form (or can't be put into that form), then this approach won't work.

Here's an outline of this approach.

Step 1 (separate). In the differential equation, put all x's on the right and all y's on the left, to get $(g(y))^{-1}\,dy = f(x)\,dx$.

Step 2 (integrate). Add integral signs to both sides, to get $\int (g(y))^{-1}\,dy = \int f(x)\,dx$. Then perform the required integration, remembering the "$+C$" on the right.

Step 3 (evaluate). Substitute the initial condition $y(x_0) = y_0$ into the result from Step 2, to evaluate (that is, solve for) the constant C. Put this value of C back into the Step 2 result.

Step 4 (solve). Solve the result of Step 3 for y.

Separation of variables strategy for initial value problems of the form (4.54)

Of course, this strategy relies on finding manageable antiderivatives of $(g(y))^{-1}$ and $f(x)$, and as we've seen, this is not always possible. But it's possible often enough for the method to be useful.

Like integration by substitution, separation of variables entails manipulation of "differentials" like dx. (See Step 1 of the above boxed strategy.) At the end of this subsection, we'll justify this kind of manipulation in the present setting, as we did in the previous section for the substitution setting.

Example 4.8.2 The differential equation $dy/dt = kt$. We know from our studies in Sect. 3.1 that the exponential function $y = y_0 e^{kt}$ is *the* solution to the initial value problem

$$\frac{dy}{dt} = ky, \qquad y(0) = y_0. \tag{4.55}$$

Let's put aside this knowledge for a moment, and rediscover this solution using our new method.

To simplify our discussions, let's assume that we need to solve this initial value problem only for positive values of y. This is a reasonable assumption, since the equation $y' = ky$ typically models growth of some quantity for which negative values are not realistic. (In the exercises, we'll look at the more general case.)

We proceed according to the above boxed strategy:

$$\frac{dy}{y} = k \, dt$$

$$\int \frac{dy}{y} = \int k \, dt$$

$$\ln(|y|) = kt + C$$

$$\ln(y) = kt + C. \tag{4.56}$$

The last step is because, again, we are assuming that $y > 0$; so $|y| = y$.

Into Eq. (4.56) we substitute the initial condition $y(0) = y_0$, to get

$$\ln(y_0) = k \cdot 0 + C = C,$$

so again by (4.56),

$$\ln(y) = kt + \ln(y_0).$$

To solve for y, we exponentiate both sides:

$$e^{\ln(y)} = e^{kt + \ln(y_0)} = e^{kt} e^{\ln(y_0)}$$

$$y = y_0 e^{kt}. \tag{4.57}$$

We have thus solved the exponential growth initial value problem "from scratch." This is quite different from what we did in Sect. 3.1—there, we *proposed* the solution $y = y_0 e^{kt}$, and then verified that it worked. Generally speaking, in "real life," one is not supplied with even a proposed solution; typically, one has to find one, as we did in the above example.

That example illustrates the separation of variables technique in an *autonomous* context, meaning, again, where a derivative is expressed solely in terms of the *dependent* variable. If on the other hand, our differential equation is *pure-time*—the derivative is expressed solely in terms of the *independent* variable—then the separation of variables method is not required; see, for instance, Examples 4.6.4 and 4.7.2(b) above. It's worth noting, though, that separation of variables *may* be applied in pure-times situations. We illustrate this in part (a) of the following example.

Example 4.8.3 Use separation of variables to solve:
(a) The initial value problem

$$\frac{dy}{dx} = 4x^3, \qquad y(-1) = 5.$$

(b) The differential equation

$$\frac{dy}{dx} = \frac{\cos(x)}{e^y};$$

(c) The initial value problem

$$\frac{ds}{dt} = e^{\sin(t)}\cos(t)\sqrt{s}, \qquad s(0) = 9.$$

Solution. (a) Applying the boxed strategy above, we have

$$dy = 4x^3\,dx$$
$$\int dy = \int 4x^3\,dx$$
$$y = x^4 + C.$$

Our initial condition then gives us $5 = y(-1) = (-1)^4 + C = 1 + C$, so $C = 5 - 1 = 4$, so

$$y = x^4 + 4.$$

(b) Our boxed procedure still applies when there is no initial condition provided, except that we ignore Step 3. Thus:

$$e^y\,dy = \cos(x)\,dx$$
$$\int e^y\,dy = \int \cos(x)\,dx$$
$$e^y = \sin(x) + C$$
$$y = \ln(\sin(x) + C).$$

(c) This example demonstrates how separation of variables can sometimes be used in conjunction with integration by substitution:

$$\frac{ds}{dt} = e^{\sin(t)} \cos(t) \sqrt{s}$$

$$s^{-1/2} \, ds = e^{\sin(t)} \cos(t) \, dt$$

$$\int s^{-1/2} \, ds = \int e^{\sin(t)} \cos(t) \, dt \qquad\qquad u = \sin(t)$$

$$\int s^{-1/2} \, ds = \int e^u \, du \qquad\qquad du = \cos(t) \, dt$$

$$2s^{1/2} = e^u + C = e^{\sin(t)} + C.$$

Putting in $s(0) = 9$ gives

$$2 \cdot 9^{1/2} = e^{\sin(0)} + C = e^0 + C = 1 + C,$$

so $C = 2 \cdot 9^{1/2} - 1 = 6 - 1 = 5$, so

$$2s^{1/2} = e^{\sin(t)} + 5$$

$$s^{1/2} = \frac{e^{\sin(t)} + 5}{2}$$

$$s = \left(\frac{e^{\sin(t)} + 5}{2} \right)^2 .$$

4.8.2   Diffusion Across a Cell Membrane

As an application of the separation of variables method, we consider a cell in some ambient environment (say, the bloodstream). Suppose that environment contains a substance (a medication, a chemical, etc.) that can pass through the cell membrane.

   Let $K(t)$ denote the concentration of the substance *inside* the cell, as a function of time. (We might measure $K(t)$ in milligrams per liter, for example, and t in minutes.) Any change in $K(t)$ is called *diffusion*. We can model diffusion with the initial value problem

$$\frac{dK}{dt} = \beta(\gamma - K), \qquad K(0) = K_0, \qquad\qquad \text{(DM)}$$

for certain *positive* parameters β and γ, and K_0. We explain:

- The quantities β, γ, and K_0 are positive constants (parameters). More specifically:

   (a) The parameter β is called the **diffusion rate**.

(b) The parameter γ represents the initial concentration of the substance **outside** the cell. (Strictly speaking, this concentration will also change with time. But in many cases, this concentration will change slowly enough (compared to the concentration inside the cell) that we may assume it is constant.

(c) The constant K_0 represents the initial concentration of the substance **inside** the cell.

- The differential equation in the initial value problem (DM) therefore tells us: The concentration K of the substance inside the cell changes at a rate proportional to the *difference* between the concentration of the substance outside the cell, and the concentration inside the cell. In particular, this differential equation tells us that, not surprisingly, the larger the difference between internal and external concentrations, the more rapid the diffusion.

Let us assume, further, that the initial concentration of the substance inside the cell is **less** than the concentration outside the cell. We may then use separation of variables to solve the above initial value problem (DM), as follows.

First we separate variables in the differential equation in (DM), and then take integrals on both sides, to get

$$\int \frac{dK}{\gamma - K} = \int \beta \, dt. \tag{4.58}$$

The integral on the right-hand side of (4.58) equals $\beta t + C$, where C is some constant. We evaluate the integral on the left hand side by substituting $u = \gamma - K$. Since γ is constant with respect to K, this gives us $du = -dK$, so that

$$\int \frac{dK}{\gamma - K} = \int \frac{-du}{u} = -\ln(|u|) + C = -\ln(|\gamma - K|) + C.$$

So (4.58) yields

$$-\ln(|\gamma - K|) = \beta t + C. \tag{4.59}$$

But we're assuming that the concentration inside the cell starts out less than that outside. It's reasonable to also assume, then, that the internal concentration will *never* exceed the external concentration. In other words, $\gamma - K > 0$ always. We may therefore remove the absolute value symbols in (4.59), so that

$$-\ln(\gamma - K) = \beta t + C. \tag{4.60}$$

Next, we plug the initial condition $K(0) = K_0$ into (4.60), to get $-\ln(\gamma - K_0) = C$, and put this value of C back into (4.60), to get

$$-\ln(\gamma - K) = \beta t - \ln(\gamma - K_0). \tag{4.61}$$

Some algebra is required to solve (4.61) for K:

- First, we multiply both sides of (4.61) by -1; we get

$$\ln(\gamma - K) = -\beta t + \ln(\gamma - K_0). \tag{4.62}$$

- We then exponentiate both sides of (4.62), yielding

$$\gamma - K = e^{-\beta t + \ln(\gamma - K_0)} = e^{-\beta t} e^{\ln(\gamma - K_0)} = (\gamma - K_0)e^{-\beta t}. \tag{4.63}$$

- We solve (4.63) for K; we get

$$K = \gamma - (\gamma - K_0)e^{-\beta t}. \tag{4.64}$$

This is, finally, our solution to the intial value problem (DM).

As a partial check on our work, we note that, if we plug $t = 0$ into (4.64), we get

$$K(0) = \gamma - (\gamma - K_0)e^{-\beta \cdot 0} = \gamma - (\gamma - K_0) = K_0,$$

agreeing with the initial condition in (DM).

We note further that, as $t \to +\infty$, $e^{-\beta t}$ approaches zero (since $\beta > 0$), and therefore, by (4.64), $K(t)$ approaches $\gamma - 0 = \gamma$. In other words (not surprisingly), as times continues to elapse, the concentration inside the cell approaches the concentration outside.

4.8.3   Justification

Each of our separation of variables arguments has begun in the following fashion:

$$\frac{dy}{dx} = f(x)g(y) \tag{a}$$

$$(g(y))^{-1} dy = f(x) dx \tag{b}$$

$$\int (g(y))^{-1} dy = \int f(x) dx. \tag{c}$$

In making such an argument, we've pretended, much as we did in integrating by substitution, that the so-called "differentials" dx and dy represent actual quantities. But again, they don't; they are just pieces of the symbol dy/dx. So why do arguments like the above work?

The answer is that the middle statement (b) is just a heuristic, written to help guide us from (a) to (c). But in fact, we don't really need (b). This is because (c) follows from (a) in another way, that does not entail the differentials that appear in (b).

Here's how: assume that (a) is true. Let F be an antiderivative of f, so that the right-hand side of equation (c) equals $F(x) + C$. If we can show that the left-hand side of (c) equals the same thing, then the two sides of (c) will be equal, and we'll be done.

So we need only show that

$$\int (g(y))^{-1} \, dy = F(x) + C,$$

which is the same, by the boxed statement in Sect. 4.6.3, as showing that

$$\frac{d}{dx} \left[\int (g(y))^{-1} \, dy \right] = F'(x). \tag{4.65}$$

We demonstrate the latter by the chain rule:

$$\frac{d}{dx} \left[\int (g(y))^{-1} \, dy \right] = \frac{d}{dy} \left[\int (g(y))^{-1} \, dy \right] \frac{dy}{dx} = (g(y))^{-1} \frac{dy}{dx} = f(x) = F'(x)$$

(the second-to-last equality is by equation (a)), and we're done.

4.8.4   Exercises

4.8.4.1   Part 1: The Separation of Variables Method

1. Use the method of separation of variables to find a formula for the solution of the differential equation $dy/dt = (y + 5)^2 e^t$. Your formula should contain an arbitrary constant to reflect the fact that many functions solve the differential equation.

2. Use the method of separation of variables to find formulas for the solutions to the following initial value problems.

   (a) $dy/dt = 1/y^2$, $\ y(0) = 1$.

   (b) $dz/dx = 3(x - 2)$, $\ z(4) = 2$.

   (c) $dy/dx = x^2/y^2$, $\ y(3) = -1$.

   (d) $dy/dx = \cos^2(y)e^x$, $y(0) = \pi/4$.
   (Hints: $1/\cos^2(y) = \sec^2(y)$, and an antiderivative of $\sec^2(y)$ is $\tan(y)$.)

   (e) $du/dv = v(1 + u^2)$, $\ u(2) = 3$.

4.8.4.2   Part 2: Other Applications and Contexts

3. **Newton's law of cooling.** According to Newton's law of cooling, in a room where the ambient temperature is A, the temperature C of a hot object will change according to the differential equation

$$\frac{dC}{dt} = -k(C - A).$$

   The constant k gives the rate at which the object cools.

   (a) Find a formula for the solution to this equation using the method of separation of variables. Your formula should contain an arbitrary constant. **Note:** the process here, and the end result, are quite similar to those of the subsection "Diffusion across a cell membrane" above.

You might want to check your result against Exercise 12(c) of Sect. 3.1.5.

(b) Suppose A is $20\,^\circ$C and k is 0.1° per minute per $^\circ$C. If time t is measured in minutes, and $C(0) = 90\,^\circ$C, what will C be after 20 min?

(c) How long does it take for the temperature to drop to $30\,^\circ$C?

4. (a) Suppose a cold drink at $36\,^\circ$F is sitting in the open air on a summer day when the temperature is $90\,^\circ$F. If the drink warms up at a rate of $0.2\,^\circ$F per minute per $^\circ$F of temperature difference, write a differential equation to model what will happen to the temperature of the drink over time.

(b) Obtain a formula for the temperature of the drink as a function of the number of minutes t that have passed since its temperature was $36\,^\circ$F.

(c) What will the temperature of the drink be after 5 min; after 10 min?

(d) How long will it take for the drink to reach $55\,^\circ$F?

5. **A leaking tank**. One may show that the differential equation

$$\frac{dV}{dt} = -k\sqrt{V}$$

models the volume $V(t)$ of water in a leaking tank, after t hours.

(a) Use the method of separation of variables to show that

$$V(t) = \frac{k^2}{4}(C - t)^2$$

is a solution to the differential equation, for any value of the constant C.

(b) Explain why the function

$$V(t) = \begin{cases} \dfrac{k^2}{4}(C - t)^2 & \text{if } 0 \le t \le C, \\ 0 & \text{if } C < t. \end{cases}$$

is *also* a solution to the differential equation. Why is *this* solution more relevant to the leaking tank problem than the solution in part (a)?

6. **A falling body with air resistance.** The differential equation

$$\frac{dv}{dt} = -g - bv$$

may be to model the motion of a body falling under the influence of gravity (g, a constant) and air resistance (bv). Here b is a positive constant, and v is the velocity of the body at time t.

(a) Solve the differential equation by separating variables, and obtain

$$v(t) = \frac{1}{b}\left(Ce^{-bt} - g\right),$$

where C is an arbitrary constant. Hint: to integrate in v, substitute $u = g + bv$, remembering that g and b are constant with respect to v.

(b) Now impose the initial condition $v(0) = 0$ (so the body starts its fall from rest) to determine the value of C. What is the formula for $v(t)$ now?

(c) The distance $x(T)$ that the body has fallen by time T is given by the integral

$$x(T) = \int_0^T v(t)\,dt, \quad \text{because} \quad \frac{dx}{dt} = v \quad \text{and} \quad x(0) = 0.$$

Use your formula for $v(t)$ from part (b) to find $x(T)$.

4.8.5   Summary

Suppose $E(t)$ and $p(t)$ are measurable quantities such that, whenever p is constant on an interval, the net change in E over that interval equals the value of p on that interval times the length Δt of the interval. (For example, E might be energy and p power, or E might be displacement and p velocity.) Then we say that E is an **accumulation function** for p, and we denote the net change in E over *any* interval $[a, b]$, whether p is constant there or not, by the **definite integral**

$$\int_a^b p(t)\,dt.$$

Such a definite integral can be evaluated, at least approximately, using the **Riemann sum** method. With this method, we pretend that p is constant "in pieces," so that, on each of these pieces, we can pretend that the change in E equals the constant value of p there times the length of the piece. We then put all of these "pieces of net change" together to approximate the overall net change, which is to say the definite integral. And the narrower the pieces, the better the approximation. In this way, we can study the cumulative effects of many phenomena.

The Fundamental Theorem of Calculus tells us that, under appropriate conditions,

$$\int_a^b f(x)\,dx = F(b) - F(a),$$

where F is an **antiderivative** of f. In other words, the Fundamental Theorem of Calculus tells us that, if we can **antidifferentiate**, then we can evaluate definite integrals more directly and exactly than were we to use the Riemann sum method. (But we *can't* always antidifferentiate. For example, consider the function $f(x) = e^{-x^2/2}$, which has no closed-form antiderivative. Or consider the function $p(t)$ of Fig. 4.3, which is not given by a formula at all. So Riemann sums are still essential.)

The process of antidifferentiation has applications beyond the study of definite integrals. Using antiderivatives and **indefinite integrals**, together with integration techniques like **substitution** and **separation of variables**, we can find closed-form solutions to a variety of **differential equations** and **initial value problems**, of **pure-time**, **autonomous**, and **mixed** type.

Printed in the United States
by Baker & Taylor Publisher Services